Lengsfeld / Mainka / Altstädt
Carbon Fibers

Hauke Lengsfeld
Hendrik Mainka
Volker Altstädt

Carbon Fibers

Production, Applications, Processing

HANSER

Hanser Publishers, Munich Hanser Publications, Cincinnati

The Authors:

Dr.-Ing. Hauke Lengsfeld, Schill+Seilacher "Struktol" GmbH, Hamburg

Dr.-Ing. Hendrik Mainka, Volkswagen Group of America

Prof. Dr.-Ing. Volker Altstädt, Neue Materialien Bayreuth GmbH

Distributed in the Americas by:
Hanser Publications
414 Walnut Street, Cincinnati, OH 45202 USA
Phone: (800) 950-8977
www.hanserpublications.com

Distributed in all other countries by:
Carl Hanser Verlag
Postfach 86 04 20, 81631 Munich, Germany
Fax: +49 (89) 98 48 09
www.hanser-fachbuch.de

Library of Congress Control Number: 2020946062

Editor: Dr. Mark Smith
Production Management: Der Buchmacher, Arthur Lenner, Windach
Coverconcept: Marc Müller-Bremer, www.rebranding.de, Munich
Coverdesign: Max Kostopoulos
Typesetting: Kösel Media, Krugzell, Germany
Printed and bound by CPI books GmbH, Leck
Printed in Germany

ISBN: 978-1-56990-828-0
E-Book ISBN: 978-1-56990-829-7

The Authors

Dr.-Ing. Hauke Lengsfeld

After completing his studies in Chemistry and his Ph.D. at the Department of Polymeric Materials at the University of Bayreuth (Germany) (Adviser: Prof. Dr.-Ing. Altstädt), Dr.-Ing. Hauke Lengsfeld was responsible for the production of large CFRP components at Airbus Operations GmbH as a materials and process expert. He was project manager for the development of CFRP components for the Airbus A350XWB. Subsequently, he was Head of Technical Marketing and Sales at Saertex Stade GmbH. At Hexcel Composites GmbH he worked as Technical Support Manager Aerospace for Central Europe. In close cooperation with other coauthors, he wrote his first book on prepreg processing, *Composite Technology*. Since 2018 he has been working as head of development of reactive polymers, epoxy products, and flame retardants at Schill+Seilacher "Struktol" GmbH. Since 2007, Dr.-Ing. Hauke Lengsfeld is also an adjunct professor, teaching courses on fiber reinforced materials and chemistry at several universities.

Dr.-Ing. Hendrik Mainka

After completing his studies in chemical engineering with a focus on molecular and structural product design at the Otto-von-Guericke University Magdeburg (Germany), Dr.-Ing. Hendrik Mainka received his Ph.D. in production engineering (Prof. Dr. Prof. Dr. Axel Herrmann – Faserinstitut Bremen e.V. (FIBRE))

Since 2011 Dr.-Ing. Mainka has been working for Volkswagen in the Group Research – Materials and Manufacturing Processes department. The focus here was on alternative fiber reinforcements and automotive lightweight construction. In 2015 he moved to Volkswagen Group of America, where he has been leading the Product Innovation since 2016. His work here focuses on lightweight automotive construction using fiber composites, electromobility, and additive manufacturing.

In January 2020 Dr.-Ing. Mainka took over the responsibility for the Volkswagen Group Innovation Hub Knoxville. With focus on material research and innovations for electric vehicles, the Knoxville Innovation Hub is an important location of the global Volkswagen Group Innovation network.

Dr.-Ing. Mainka also worked for a year at the Oak Ridge National Laboratory, where he was involved in the development of cost-effective automotive carbon fibers.

Prof. Dr.-Ing. Volker Altstädt

Volker Altstädt undertook university studies in physics and received a Dr.-Ing. in 1987 with Prof. Dr.-Ing. Ehrenstein at the Department of Mechanical Engineering at the University of Kassel (Germany). For eight years he was working at the Department for Polymer Composites and later for Polymer Physics at BASF AG in Ludwigshafen. Since 1995 he has been professor for Plastics in Mechanical Engineering and Head of the Department of Polymers and Plastics Composites at the Technical University of Hamburg-Harburg. Since October 2000 he has been full professor and Director of the Department of Polymer Engineering at the University of Bayreuth. In summer 2009 he became in addition the CEO of the Bavarian Research Company NMB GmbH in Bayreuth.

Preface

According to the composite market report of CCeV and AVK, a global carbon fiber market of about 70.5 thousand tons was identified for 2017 [1]. Mordor Intelligence expects an annual growth rate of 13.45% for carbon fibers for the period 2018 to 2023. Compared to other fibers such as glass fibers with several million production tons/year at a growth rate of approx. 7.4% [3], the consumption of carbon fibers worldwide is very low but the volume growth is very high. On the other hand, the applications are highly innovative and mostly driven by the goal of weight reduction. Demand is highest in the aerospace industry (36%) and the automotive industry (24%). But also the wind energy and sports industries with 13% each and the construction industry (5%) are important sales markets [2].

Against this background, the aim of this technical book is to present for the first time a book on fiber composite technology with a focus on modern carbon fibers. The focus is on the individual characteristics of carbon fibers, from the precursor for the fiber, the stabilization and carbonization process, to the resulting atomic structure and the properties. The various paths from the fiber to the roving to the composite material and its properties are discussed using many current examples from various branches of industry.

After a brief introduction to the historical development of carbon fibers, the *second chapter* of the book first explains the atomic model for carbon compounds in order to make the various carbon modifications (diamond, C-nanofiber, fullerenes …) comprehensible to the reader.

The *third chapter* describes in great detail the process steps of precursor production, oxidative stabilization, and carbonization/graphitization. The discussion of surface treatment and the correct sizing of the fibers is also important for the reader. Very instructive are the many graphic representations, e.g. of C-fibers with and without sizing, as well as the ideas about the interphase.

This chapter is logically followed in *Chapter 4* with the analysis of carbon fibers and rovings made from them. In addition to a presentation of the relevant standards, current test procedures for characterization and classification of the processing behavior of carbon fiber yarns and filaments are addressed. Test apparatuses are clearly illustrated in pictures and, for example, drapery faults caused by forming are clearly explained by means of photos.

A particular strength of the book is the presentation of the applications of carbon fibers in *Chapter 5*. Due to the excellent and diverse properties of PAN-based fibers, they are used in a wide variety of applications. The book gives application examples of the different fiber types (PREOX and carbon fibers) and explains with examples the different presentations (continuous, cut, ground) from a practical perspective and with rich illustration, to a degree that can hardly be found in other sources. The presentation is subdivided according to the different fiber types and covers the broad spectrum of production methods for applications in the aerospace industry, from automotive structures made of CFRP based on multiaxial fabrics to PrimeTex fabrics for winter sports skis, carbon short fibers for screed concrete or in a 3D-printed fan housing, to name just a few examples.

Since C-fiber production is very energy-intensive, it is no longer possible to talk about materials today without addressing their recycling and carbon footprint in the lifecycle analysis. *Chapter 6* is therefore devoted to a current review of C-fiber recycling and sustainability. Here, too, a distinction is made between the different presentations (continuous, cut, ground) in the description of the various recycling paths. In addition to the actual recycling cycle for C-fibers, the challenges of waste treatment are also addressed and ways are shown of moving from a material mix of different residual materials as an input variable to an energetic or better recyclate semi-finished product (flour, balls, or granulate) by thermal, chemical, or mechanical processes, which lead to sustainable secondary use in a variety of applications shown in the book. The authors come to the conclusion that the recycling of all types of carbon-containing waste and the associated recovery of high-quality carbon fibers (rCF) has been solved on an industrial scale and is increasingly applied.

Nevertheless, there are still challenges to be met in order to be able to expand the use of carbon fibers. These are mainly driven by the high price of the fiber. In *Chapter 7*, precursor costs, energy costs, and investment costs for plants are subdivided into future improvement potentials.

In summary, this technical book can be used both as a manual and as a textbook or reference book for practitioners. The physical basics of carbon fibers, the manufacturing processes, their material, and process engineering basics are taught with a strong practical orientation. The book also aims to make a contribution at the interfaces between carbon fiber technology and semi-finished products and components made from them, and to build the necessary bridges between structure/production and properties. Looking to the future, the book presents important developments in the field of recycling and energy efficiency of modern carbon fiber composite technology.

Bayreuth, February 2019 *Volker Altstädt*

References for the Preface

[1] Der globale CF- und CC-Markt – Michael Sauer, Michael Kühnel (CCeV) Der GFK-Markt Europa – Dr. Elmar Witten (AVK)

[2] Study of the Carbon Global Fiber Market – *https://www.mordorintelligence.com/industry-reports/global-carbon-fiber-market-industry*

[3] *https://www.reportlinker.com/p05176097/Global-Glass-Fiber-Market.html*

Acknowledgements

We would like to thank everyone who supported us in the production of this book.

We would like to thank Carl Hanser Verlag for the opportunity to publish this book, especially Ms. Ulrike Wittmann for the design and formatting of the chapters.

We personally thank Roland Bernicke, Tejin Carbon Europe, Jörg Heubischl, AFF, Thomas Neumeyer, Neue Materialien Bayreuth, Daniel Stelzl, MT Aerospace, as well as the companies Karl-Meyer AG, Schunk Kohlenstofftechnik GmbH, FACC AG, Airbus Operations GmbH, Hexcel Composites GmbH, National Oak Ridge Labs, and many others for the pertinent hints, stimulating discussions, and pictures.

Special thanks go to Tejin Carbon Europe GmbH, Textechno GmbH, Audi AG, and Brose Fahrzeugteile GmbH & Co, who provided an important part of the images.

We would also like to thank our families for their patience.

Abbreviations

AFP	automated fiber placement
AM fiber	advanced modulus fiber
ATL	automatic tape laying
BMC	bulk molding compound
CF	carbon fiber
CFC	carbon fiber reinforced carbon
CFRP	carbon fiber reinforced plastic
CPT	cured ply thickness
CSiC	carbon fiber reinforced silicon carbide
CVD	chemical vapor deposition
DFP	dry fiber placement
FAW	fiber aerial weight
Filaments	fine, individual fibers in a yarn
FMA	fiber matrix adhesion
FVC	fiber volume content
HT fiber	high tensile fiber
IM fiber	intermediate modulus fiber
K number	number of filaments in a yarn bundle in 1000s
MPP	mesophase pitch
NCF	non-crimped fabrics, see MAG (multiaxial fabrics)
PA	polyamide
PAN	polyacrylonitrile
PEEK	polyether ether ketone
PES	polyethersulfone
PREOX	pre-oxidized fiber
prepreg	PREimPREGnated fibers
rCF	recycled carbon fiber
RTM6	aerospace qualified epoxy resin system
SMC	sheet molding compound

tow	bundle of fiber filaments, *see also* yarn
UD	unidirectional
UHM fiber	ultra high modulus fiber
yarn	bundle of fiber filaments, *see also* tow
XPS	X-ray photoelectron spectroscopy

Contents

1 Introduction

Uniqueness

The uniqueness of carbon fibers is defined by their excellent properties. Their strength is comparable to that of high-performance steel and their stiffness is higher than that of all known metals, ceramics, or polymers. Even their thermal and electrical conductivity exceeds that of comparable materials (see Chapter 3). The uniqueness of the properties of carbon fibers becomes apparent when the strength or stiffness is divided by the material density to obtain the specific properties. No other known material can currently compete with this result [1-1].

History

Carbon fibers are a synthetic product. The history of carbon fiber began in 1879 with the invention of the first filament lamp by *Thomas Edison*. Edison experimented with fine cotton and bamboo fibers in order to improve the already known arc lamp. He pyrolyzed (carbonized) the fibers at high temperatures. The resulting carbon filaments showed good electrical conductivity and stability for use in incandescent lamps. Unfortunately this application for carbon fiber was soon replaced by tungsten filaments for this application due to their easier and more cost effective manufacturing process. As the result of this development, the initially interest in carbon fiber slowed down again [1-2].

The modern era of carbon fibers did not begin until the 1950s and 1960s. Fundamental development work was done on characterizing the first carbon fibers based on polyacrylonitrile (PAN), pitch, viscose, and rayon in the development laboratories of *Union Carbide* and *DuPont*. It was discovered that pyrolysis of carbon-rich materials resulted in an oriented graphite structure, with a very high modulus of elasticity and strength values. This discovery could be demonstrated in practical tests on so-called graphite whiskers. Whiskers are thread-shaped single crystals with an anisotropic lattice structure [1-3]. Soon several researchers independently recognized the necessity of stretching the fiber at high temperatures (> 2800 °C) in order to further improve the properties, especially the modulus of elasticity. This allowed up to 10 times higher stiffnesses to be achieved than without the stretching process. One of the first commercial fibers from Union Carbide was the "Thornel-25". This trade name derives from the Nordic deity *Thor* ("strength").

At the beginning of the 1960s, Japanese scientists were able to successfully demonstrate in a study the production of high modulus or high stiffness fibers based on polyacrylonitrile (PAN) as a precursor. Earlier attempts of US-American scientists had failed here. The secret of this development was improved PAN precursors. The chemical structure (the so-called backbone) after carbonization was the reason for their excellent properties. At the same time, British scientists also succeeded in developing high-modulus PAN-based fibers, which were soon commercialized. All these developments created the basis for the carbon fibers used today. Through increasing application and further development, production costs have been drastically reduced in the past 35 years through new processes and new raw materials [1-4].

Since around 2010, research and subsequent commercialization have focused on new precursors such as lignin and textile PAN on the one hand, as well as measures for energy efficiency on the other (plasma oxidation, microwave oven) [1-5]. The commercialization of these new technologies is driven by innovative start-ups such as *LeMond Composites, RMX,* and *4M* [1-6] [1-7] [1-8].

Current Applications

Due to their outstanding properties, carbon fibers are used in a wide variety of industrial sectors. Examples range from silicon carbide brake disks for aircraft and sports cars, tanks for solid-fuel rockets for space travel, hydrogen tanks and structural components for automobiles, to applications in the construction industry and sporting goods. These and other examples are discussed in detail in Chapter 5.

Recycling

Waste containing carbon fibers occurs in various stages during the production, processing, and use of carbon fibers and carbon fiber reinforced polymer (CFRP) components made from them. Recycling processes include pyrolysis, fluidized bed, solvolysis, and direct chemical use. The challenges that arise and the benefits that can be achieved are described in detail in Chapter 6.

Future Trends

Future applications of carbon fibers will be in the area of additive manufacturing (e.g., Big Area Additive Manufacturing), but also widespread applications in automotive and consumer-oriented areas seem possible if the prices for carbon fibers continue to fall as predicted. The most important requirement for this will be the successful commercialization of new precursors and conversion technologies for the production of carbon fibers.

References for Chapter 1

[1-1] *Handbook of Composites*, 2nd ed., edited by S.T. Peters, Chapman & Hall, 1998, ISBN 0-412-54020-7

[1-2] T. Edison; US 470925 1880-03-15

[1-3] *Faserverbundbauweisen*, M. Flemming, G. Ziegmann, S. Roth, Springer Verlag, 1995, ISBN3-540-58645-8

[1-4] *High Performance Carbon Fibers*, National Historic Chemical Landmarks of American Chemical Society, September 2003, *https://www.acs.org/content/acs/en/education/whatischemistry/landmarks/carbonfibers.html*

[1-5] H. Mainka, O. Täger, E. Körner, L. Hilfert, S. Busse, F.T. Edelmann, A.S. Herrmann, *Lignin - an alternative precursor for sustainable and cost-effective automotive carbon fiber*; Journal of Materials Research and Technology; Volume 4, Issue 3, July–September 2015, pp. 283–296

[1-6] LeMond: *https://lemondcarbon.com/*, accessed February 2019

[1-7] RMX Technologies: *http://www.rmxtechnologies.net/*, accessed February 2019

[1-8] 4M: *http://4mio.com/*, accessed February 2019

2 Carbon and Its Properties

This chapter will provide an overview concerning the chemical and physical properties of carbon. It explains the structure of the carbon atom and its atomic model, as well as the three possible modifications of carbon. A short overview of the different structural forms (allotropes) of elemental carbon links the chemical and physical properties of carbon at the molecular level with the macroscopic properties of each allotrope.

In the periodic table, carbon can be found at position number 6 with a molar mass of 12.011 g/mol. Carbon has seven isotopes. ^{12}C at 98.89% and ^{13}C at 1.108% are the two most common and only stable isotopes. The isotopes ^{10}C, ^{11}C, ^{14}C, ^{15}C, and ^{16}C are radioactive. That means they decay, with emission of β radiation, which can be either negative (electron) or positive (positron) [2-1].

Carbon occurs in nature both in pure elemental form (e.g., diamond, graphite, fullerenes) and chemically bound (e.g., in the form of carbides, carbonates, carbon dioxide, crude oil, natural gas, and coal). Due to its special electron configuration (Section 2.1.2), carbon has the capability to form complex molecules and has the largest variety of chemical compounds of all chemical elements. This complexity is also reflected in the allotropy of carbon with its multitude of compounds and their specific properties (Section 2.3). For example, diamond is the hardest known material, whereas graphite is relatively soft.

Compounds of carbon and hydrogen form the basis of organic chemistry with a variety of over six million known compounds [2-2].

2.1 Structure of the Carbon Atom and Atomic Model of Carbon

Throughout history, the atomic model of carbon has been constantly improved and expanded. The spectrum ranges from the *Thomson* atomic model, according to which the atom consists of uniformly distributed mass and positive charge in which the electrons move, to the hybrid orbital model, which was derived mathematically from a linear combination of the wave functions of the basic atomic orbitals.

In this chapter the *Rutherford* and *Bohr* atomic model and the hybrid orbital model are briefly explained in order to make the possible modifications of carbon more understandable.

The Rutherford atomic model is an atomic model that was developed by Ernest Rutherford in 1911. It corrects the Thomson atomic model that was previously taken to be valid.

Bohr's atomic model is based on Rutherford's atomic model and forms the basis for the atomic orbitals and hybridization of carbon, which are explained in Section 2.1.2.

2.1.1 Rutherford's and Bohr's Atomic Models

Using his scattering formula, Rutherford calculated values for the size and charge distribution of various elements based on experiments. His conclusion was that the enormous electric field strength required for the measured strong deflection of alpha particles can only be explained under the assumption that the positive charge of the atom is completely concentrated in a compact atomic nucleus, whose radius is a factor of about 3000 smaller than the atomic radius.

To explain the electrical neutrality of atoms, Rutherford assumed that the atomic nucleus is surrounded by electrons, with the total number of electrons per atom corresponding exactly to the number of nuclear charges. However, no information about the spatial distribution of the electrons could be derived from the experiments, since the electrons, due to their small mass, do not contribute to the deflection of the alpha rays to a detectable extent [2-3].

Rutherford did not develop his own model of the electronic structure of atoms; he merely quoted *Nagaoka's* "Saturn model of the atom" to validate his estimate of the atomic charge number [2-4].

The physical properties of the subatomic particles are summarized in Table 2.1 .

Table 2.1 Physical Properties of Subatomic Particles [2-5]

Name	Mass in kg	Relative Charge	Charge in C
proton	1.673×10^{-27}	+1	$+1.6 \times 10^{-19}$
neutron	1.675×10^{-27}	0	0
electron	0.911×10^{-31}	−1	-1.6×10^{-19}

A problem of Rutherford's atomic model is the explanation of emission and absorption of energy quanta: with Rutherford's model no explanation can be given for the so-called spectral lines of various gases. It was therefore replaced by the Bohr atomic model, which assigns different energy levels to the electrons.

Bohr formulated a new model by adding three postulates to the Rutherford model [2-6][2-7]:

- Electrons move on stable orbits around the atomic nucleus. Contrary to what the theory of electrodynamics predicts, the electrons do not emit energy in the form of electromagnetic radiation during movement in the stable orbit.
- The radius of the electron orbit does not change continuously – it changes abruptly. In this quantum leap, electromagnetic radiation is emitted (or absorbed) whose frequency results from the correlation between energy and frequency of light discovered by *Max Planck*.
- Electron orbits are only stable if the orbital angular momentum of the electron is an integer multiple of Planck's quantum of action.

Figure 2.1 illustrates the Bohr atomic model of carbon:

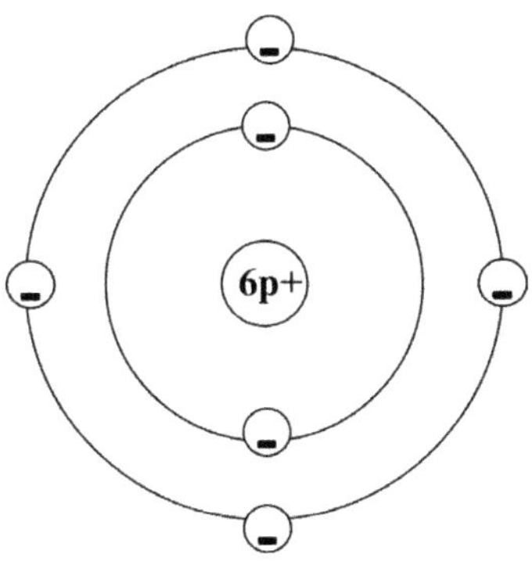

Figure 2.1
The *Bohr* atomic model of carbon

2.1.2 Atomic Orbitals and Hybridization

Orbitals are single electron wave functions (mostly abbreviated with ψ) in quantum mechanics. The absolute square of a wave function is interpreted as the probability of the electron to stay in a certain volume. In the orbital model there are no orbits like in the atomic model of Niels Bohr. The quantum mechanics led to the realization that the exact location of the electrons cannot be described stochastically due to *Werner Heisenberg's* uncertainty relation. That means only the distribution of the electrons can be described.

Since the probability of the electrons being located at a distance from the atomic nucleus is asymptotically close to zero and extends to infinity, the orbital chosen is the volume in which the electron in question is located with a probability of about 90%. This volume correspond approximately to the size of the atoms. The distances of the greatest probabilities within the orbitals of an electron correspond to the orbital distances calculated by Niels Bohr.

Orbitals are classified by the four quantum numbers n, l, m_l, and s [2–8]:

- *n (principal quantum number, range of values: n = 1, 2, 3, ...)* describes the main energy level of an electron. To a certain extent, it corresponds to the shell or orbit n of the *Bohr* atomic model.
- *l (azimuthal quantum number, orbital angular momentum quantum number, value range: l = 0, 1, ..., (n − 1))* describes the orbital angular momentum of the electron.
- *m_l (magnetic quantum number, value range: $m_l = -l, -(l-1), \dots 0, \dots +(l-1), +l$)* describes the spatial orientation of the orbital with respect to an external magnetic field.
- *s (spin quantum number, s = +½ or s = −½)* is interpreted as the spin (½: 720-degree symmetry) of the electrons.

The orbitals corresponding to the various azimuthal quantum numbers have characteristic (rough) shapes that remain qualitatively intact even at higher n values. For historical reasons, each azimuthal quantum number is given a specific letter:

Table 2.2 Characteristic Shapes of Atomic Orbitals

Name	Full Description (Historical)	Azimuthal Quantum Number	Appearance
s-orbital	sharp	l = 0	radial symmetric
p-orbital	principal	l = 1	dumbbell-shaped in the three spatial axes
d-orbital	diffuse	l = 2	crossed double dumbbell
f-orbital	fundamental	l = 3	rosette-shaped

Strictly speaking, the orbitals characterize only the possible energy state of the electron waves as they occur in single-electron systems, such as the hydrogen atom H or helium ion He^+, etc. The wave functions belonging to the orbitals result from the stationary *Schrödinger* equation of a single-electron system. Despite this limitation, however, knowledge of the rough form of the orbitals, which is also preserved in multi-electron systems, is sufficient to answer many qualitative questions about the structure of substances [2-9].

Some symmetries of chemical bonds seem to contradict the characteristic forms of orbitals. These binding symmetries only become understandable through the formation of hybrid orbitals. These are orbitals that occur in multi-particle wave functions.

A hybrid orbital is an orbital that is created mathematically from a linear combination of the wave functions of the basic atomic orbitals. This modeling process is called hybridization of the orbitals. The concept was developed by *Linus Pauling* around 1931 [2-10].

The orbitals generated by the calculation of the residence probability of the respective electrons do not always correspond to the forms assumed on the basis of chemical properties. Thus, it is found that the outer electron shell in the carbon atom has two s- and two p-electrons. Accordingly, these orbitals should lead to different types of C-H bonds in methane (CH_4). In fact, however, we find that the four bonds are identical and indistinguishable. This can be explained by sp^3 hybridization: the doubly occupied, spherical s-orbital is combined with the dumbbell-shaped p-orbitals (two singly occupied, one unoccupied) to become four identical, club-shaped sp^3-hybrid orbitals, each occupied by one electron. These are tetrahedrally aligned in space and form similar atomic bonds with the s-electrons of hydrogen. The resulting model corresponds to the observed properties of methane [2-11].

2.2 Modifications of Carbon

The model of atomic and molecular orbitals illustrates how the different structural forms of carbon occur.

Carbon has six electrons. According to the shell model (cf. *Bohr atomic model,* see Sectio .1), two electrons occupy the inner 1s shell. The 2s-level of the second shell also accommodates two electrons, and two more are contained in the $2p_x$- and $2p_y$-level. Only the four outer electrons of the second shell are chemically relevant. The probability distribution for the location of the electrons within an s-level has a spherical shape. In a p-level it is anisotropic or is often called dumbbell-shaped. The electrons occupy a dumbbell-shaped space, one dumbbell half each left and right of the center along the x-axis, if one imagines the atom placed in the center of a Cartesian coordinate system. The p_y- and p_z-orbitals (according to y- and z-axis) are perpendicular to it. Figure 2.2 schematically illustrates the s and p orbitals of carbon.

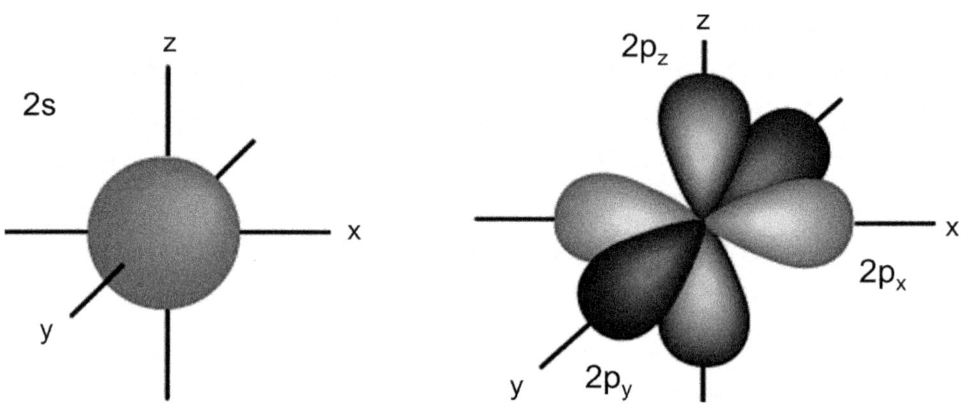

Figure 2.2 Atomic orbitals – spatial representation of 2s and 2p orbitals [2–2]

As already stated, the carbon atom has six electrons. The first two electrons occupy the 1s orbital with opposite spin (energy principle, *Pauli* principle), the next two similarly occupy the 2s orbital. The remaining two electrons occupy the p-orbitals. According to *Hund's rule*, two p-orbitals are singly occupied by these electrons.

Electrons in the s-orbitals of a shell have lower energies than those in p-orbitals (Figure 2.3).

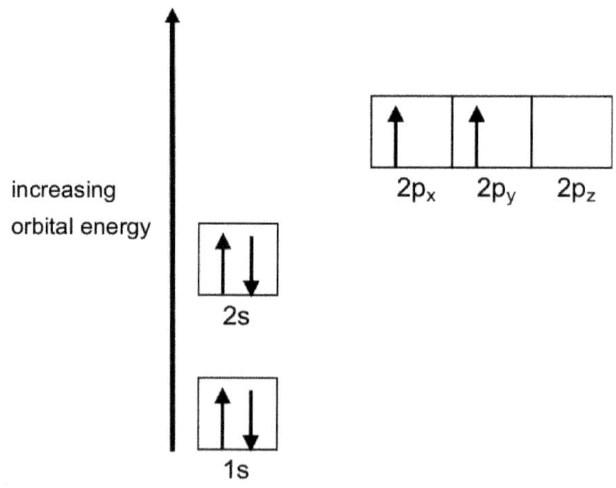

Figure 2.3 Electron occupation of the carbon atom in the ground state [2–2]

In the ground state of the carbon atom, only the two unpaired electrons of the p-orbital would be available as valence electrons. However, carbon always occurs in its compounds with four bonds, which can be clearly seen in the example of methane CH_4 (cf. Section 2.1.2).

Thus, one could assume that, taking into account Hund's rule, an electron of the s orbital is raised to the level of a p-orbital electron, which, however, would provide two different "types" of valence electrons for binding. This hypothetical thought model shows is represented in Figure 2.4:

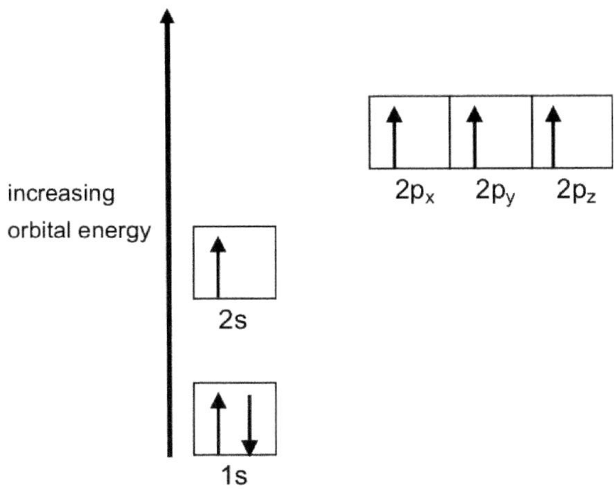

Figure 2.4 Energized state of the carbon atom [2-2]

Since an energy of about 400 kJ/mol is required to lift an electron from the s- to the p-orbital, i. e. the atom has a higher energy than in its ground state, we speak of the "excited state" of the carbon atom.

Now the carbon atom would have four binding electrons at its disposal. However, these bonds would differ energetically and in their degree of overlap of the orbitals, i. e. their bond length to other atoms in molecules.

During the molecular formation of methane from its elements, for example, one of the two 2s electrons is lifted into the unoccupied 2p-orbital level by energy input, whereby four individual electrons are now unpaired in four individual orbitals. The energy of about 400 kJ/mol required for this process is overcompensated by the energy released by the bond with four hydrogen atoms. Now it comes to a mixture of the single s-orbital with all three p-orbitals, which leads to four equivalent mixed orbitals.

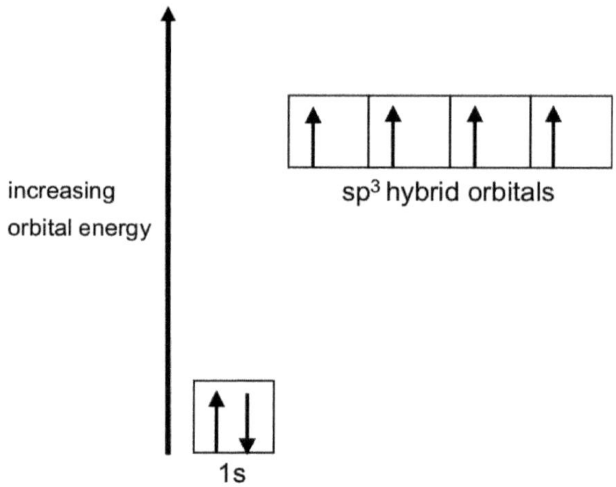

Figure 2.5 sp^3-hybrid orbitals of the carbon atom [2-2]

These orbitals are called hybrid orbitals and are responsible for the different modifications and structural forms of carbon, which are explained below.

■ 2.3 Structural Forms of Carbon

Diamond and graphite are the two main elemental structural forms (allotropes) of carbon as infinite solid carbon molecules.

In diamond, each carbon atom is sp^3-hybridized and surrounded by four more carbon atoms. All bonds in the diamond are covalent bonds and have a bond length of 0.154 nm. Due to the sp^3-hybridization, the basic shape of the diamond structure is a tetrahedron in which the carbon atoms are arranged.

In graphite, the carbon is sp^2-hybridized, resulting in a planar structure in which each carbon atom is connected to three other carbon atoms. The three covalent bonds have a bond length of 0.1415 nm. The individual layers of graphite are relatively weakly bound to each other. This attraction is called *van der Waals* forces.

In 1985 a new form of carbon was discovered with the fullerenes, which contain just sp^2-hybridized carbon. The relationship between the individual structural forms of carbon is summarized in Figure 2.6 [2-14].

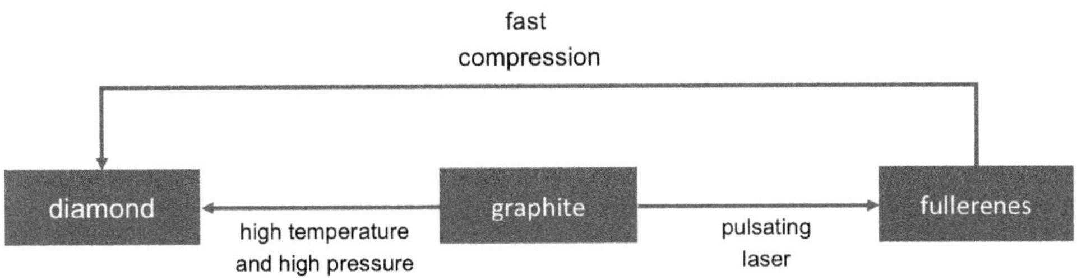

Figure 2.6 Allotropes of carbon and their interconversion

2.3.1 Diamond – The Diamond Structure (sp^3)

In diamonds, the carbon atoms are linked three-dimensionally by covalent bonds. Each carbon atom is tetrahedrally surrounded by four neighboring carbon atoms. It is bound to its neighbors by four σ bonds. The C–C single bonds (length: 1.5445 Å) are formed by overlapping tetrahedrally aligned sp^3-hybrid orbitals; see Figure 2.7.

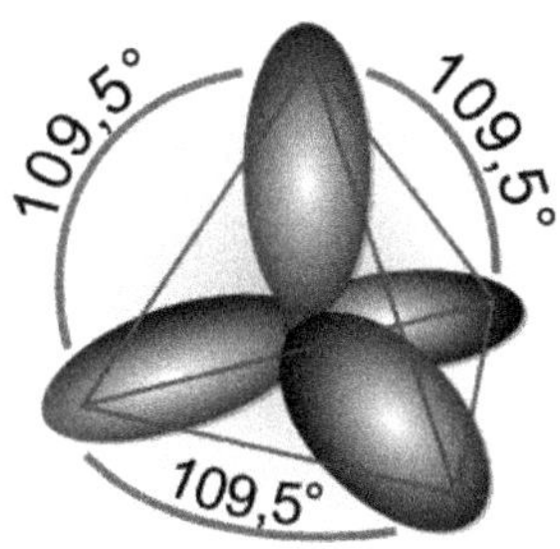

Figure 2.7
Four sp^3-orbitals – tetrahedral at equal angles [2-12]

2.3.2 Graphite – The Graphite Structure (sp^2)

Graphite is composed of flat carbon layers stacked on top of each other.

Within the layers, each carbon atom is surrounded by three carbon atoms in the form of an equilateral triangle. Each carbon atom is sp^2-hybridized and with three of its four outer electrons forms three localized σ bonds to its three atomic neighbors. The fourth valence electrons of the carbon atoms are located in delocalized π molecular orbitals, which result from a combination of the p atomic orbitals of the carbon atoms, which are not involved in hybridization and are perpendicular to the sp^2-hybrid orbitals. Thus, the carbon atoms of graphite are linked by σ bonds as well as by π bonds. In graphite, all C–C bonds have the same length.

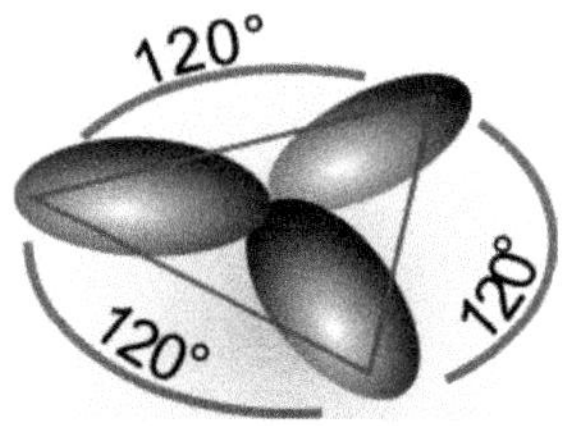

Figure 2.8
Three sp^2-orbitals are symmetrical in a plane [2–13]

2.3.3 Triple Bonding (sp^1)

If only one p-orbital hybridizes with the s-orbital, two linearly arranged π bonding lobes result. These are oriented along the x-axis; the remaining p-orbitals lie on the y- and z-axes. Two sp^1-hybridized atoms can form a carbon triple bond. An example is in the gas acetylene (ethyne), HC≡CH. Whereas bonds from sp^3-orbitals form three-dimensional structures and bonds from sp^2-orbitals form two-dimensional ones, bonds from sp^1-orbitals form at most one-dimensional (linear) chains, as in the case of acetylene H–C≡C–H, for example.

2.3.4 Other Forms of Carbon

Carbon exists in the three basic geometries presented, based on the bond structures sp^3 and sp^2 (e.g., in diamond and graphite, respectively) and sp^1 with triple bonding. In addition to these three pure geometries, there are combinations of thereof and other different forms of elemental carbon.

2.3.4.1 Fullerenes

A regular six-carbon ring, as formed by the carbon atoms in the planes of graphite, is planar. If some hexagons are replaced by pentagons, curved surfaces are formed, which form closed bodies with certain relative arrangements of the pentagonal and hexagonal rings. Such structures are realized in the fullerenes. The sp^2-bonds are no longer in one plane, but form a spatially closed structure. The smallest possible structure consists only of pentagons and requires 20 carbon atoms. One of the most stable fullerenes consists of 60 carbon atoms and contains only hexagons and pentagons that do not have a common edge with any other pentagon. The resulting pattern resembles the pattern on an (old-fashioned) football. It is called buckminsterfullerene in honor of *Richard Buckminster* [2–15].

2.3.4.2 Carbon Nanotubes

Another form of carbon comprises cylindrically arranged, sp^2-bound carbon atoms. Their geometry is formed from a flat layer of graphite that is rolled up into a cylinder. The resulting tube can also be twisted. Several single-walled tubes can lie concentrically within each other, so called multiwalled carbon nanotubes (MWCNT), in contrast to single-walled carbon nanotubes (SWCNT) [2-16].

2.3.4.3 Carbon Black

Carbon black is a black, powdery solid consisting of 80 to 99.5% carbon. Carbon black is an important technical product produced in large quantities by incomplete combustion or pyrolysis of hydrocarbons.

More than 90% of industrial carbon black is used as a filler in the rubber industry, about 70% of it for car tires and about 20% for technical rubber articles such as conveyor belts, V-belts, hoses, damping elements. The specific surface area of soot particles is about 10 to 1000 m^2/g. Nano-structuring, for example, has made it possible to specifically optimize the three most important parameters of car tires (rolling resistance, wet slip resistance, and abrasion) [2-17].

2.3.4.4 Activated Carbon

Activated carbon is a fine-grained carbon with a large inner surface, which is used as an adsorbent in chemistry, medicine, water, and waste water treatment.

Activated carbon consists mainly of carbon (usually > 90%) with a highly porous structure. The pores are interconnected like a sponge (open-pored). The inner surface is between 300 and 2000 m^2/g carbon. The density of activated carbon ranges from 0.2 to 0.6 g/cm^3 [2-1].

2.3.4.5 Carbon Fibers

Carbon fibers can be made from polymer-based precursor material. Precursor examples are polyacrylonitrile, cellulose, pitch, and lignin. These special fibers and their exact production are discussed in Chapter 3.

Carbon fibers consist mainly of graphite-like sp^2-bonded carbon with small amounts of sp^3-bonded carbon. By stretching during production it is possible to align the graphitic carbon along the fiber axis. Thus, extremely high-strength fibers with properties that approach the theoretical values of graphite along the graphite plane can be obtained. The excellent strength properties and modulus of elasticity make carbon fibers the material of choice for many applications.

Outlook

Properties and types of carbon fibers are explained in more detail in Chapter 3; test methods and applications will follow in Chapters 4 and 5.

References for Chapter 2

[2-1] A.F. Hollemann, E. Wiberg, N. Wiberg: *Lehrbuch der Anorganischen Chemie.* 102nd ed. de Gruyter, Berlin 2007, ISBN 978-3-11-017770-1

[2-2] K.P.C. Vollhardt, N.E. Schore: *Organische Chemie*, 5th ed. Wiley-VCH, Weinheim 2011, ISBN 978-3-527-32754-6

[2-3] E. Rutherford, *The Scattering of α and β Particles by Matter and the Structure of the Atom,* Phil. Mag. 6, Vol. 21, pp. 669–688 (1911)

[2-4] E. Nagaoka, *Kinetics of a system of particles illustrating the line and the band spectrum and the phenomena of radioactivity,* Phil. Mag. 7, pp. 445–455 (1904)

[2-5] J.C. Polkinghorne: *The Particle Play: An Account of the Ultimate Constituents of Matter,* 4th ed., W.H. Freeman, 1979, ISBN 9780716711773

[2-6] N. Bohr: *On the Constitution of Atoms and Molecules*, Phil. Mag. 26, pp. 1–4 (1913)

[2-7] N. Bohr: *Über die Anwendung der Quantentheorie auf den Atombau,* Zeitschr. f. Physik, 13, pp. 117–165 (1923)

[2-8] W. Heisenberg, *Physikalische Prinzipien der Quantentheorie.* 1st ed. S. Hirzel, Leipzig 1930 ISBN 978-3-41-100001-2

[2-9] E. Schrödinger: *Quantisierung als Eigenwertproblem.* In: Annalen der Physik. Vol. 79, 1926, pp. 361, 489, 734, and Vol. 81, 1926, p. 109

[2-10] L. Pauling: *General Chemistry.* W.H. Freeman, San Francisco 1949, ISBN 0-486-65622-5

[2-11] Erwin Riedel: *Anorganische Chemie.* 9th ed. Walter de Gruyter, New York 2015, ISBN: 978-3110355260

[2-12] Wikimedia Commons: *https://commons.wikimedia.org/wiki/File:Sp3-Orbital.svg#/media/File:Sp3-Orbital.svg,* accessed February 2019

[2-13] Wikimedia Commons: *https://commons.wikimedia.org/wiki/File:Sp2-Orbital.svg#/media/File:Sp2-Orbital.svg,* accessed February 2019

[2-14] R.F. Curl, R.E. Smalley, *Fullerenes*, Scientific American 265, 54–63, 1991

[2-15] H.W. Kroto, J.R. Heath, S.C. O'Brien, R.F. Curl, R.E. Smalley, *C60: Buckminsterfullerene;* Nature Vol. 318, pp. 162, 163 (November 14, 1985)

[2-16] M.-F. Yu, O. Lourie, M.J. Dyer, K. Moloni, T.F. Kelly, R.S. Ruoff: *Strength and Breaking Mechanism of Multiwalled Carbon Nanotubes Under Tensile Load.* In: Science Jan 28, 2000: Vol. 287, Issue 5453, pp. 637–640

[2-17] F. Röthemeyer, F. Sommer: *Kautschuktechnologie.* 2nd ed. Carl Hanser Verlag, Munich/Vienna 2006, ISBN 3-446-40480-5, pp. 246, 247

3 Carbon Fibers

A carbon fiber is an endlessly long, very thin material strand with a diameter of approximately 5 to 10 μm, essentially consisting of pure carbon. The carbon atoms are arranged in a graphite structure largely parallel to the fiber axis, which makes for the high stability of the fiber [3-1]. Taken together, many thousands of these carbon filaments form a fiber bundle, also known as roving or tow.

The base and thus the starting material of each carbon fiber is its precursor. Since the early work of *Thomas Edison* and the use of cellulose-based carbon fibers in light bulbs, a variety of precursors for the production of carbon fibers have been tested [3-2]. Today there are two commercially available precursors: *polyacrylonitrile (PAN)* and *pitch*.

- PAN-based carbon fibers dominate the market and represent more than 90% of the world market. This is mainly due to the good availability of the precursor PAN and its constant quality.
- The remaining approximately 10% of the world market is occupied by pitch-based carbon fibers. In addition to the classical PAN- and pitch-based precursors, there are some promising new approaches to obtain suitable starting compounds on a different chemical basis (see Chapter 7).

Heat treatment is the conventional process by which carbon fibers are produced today. Figure 3.1 schematically shows the multi-stage production process for carbon fibers based on polyacrylonitrile – a process that contains both chemical and mechanical (stretching of the fibers) components. Three characteristic process steps are performed:

1. Precursor manufacturing
2. Oxidative stabilization
3. Carbonization/graphitization

The scheme gives an idea of how complex and cost-intensive the production of carbon fibers is, especially through energy-intensive heat treatment. This is a quasi-continuous process with several weeks of production time for one production lot.

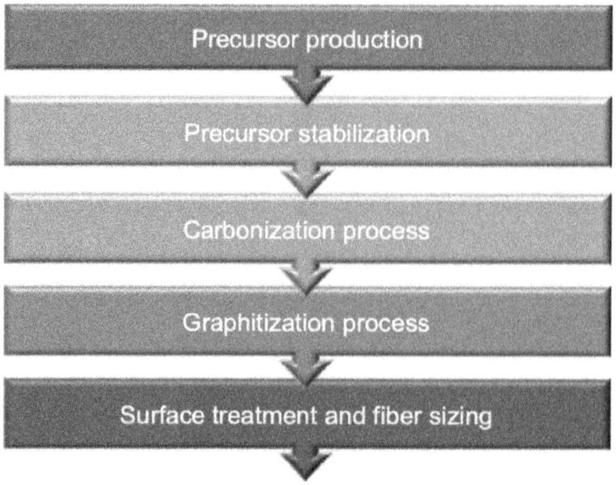

Figure 3.1
Schematic diagram of the manufacturing process for carbon fibers

After the first carbonization, so-called standard fibers or high-strength fibers (HS and HT types) are obtained. For high modulus fibers (IM, HM, UHM types) another step follows, graphitization. At the end of the production process, the carbon fiber obtained consists of > 95% pure carbon.

■ 3.1 Precursor Manufacturing

Suitable precursors are a fundamental prerequisite for the production of carbon fibers. These are long-chain, organic polymers with characteristic properties that significantly determine the later properties of the carbon fiber obtained from them. The exact chemical composition of precursors varies from manufacturer to manufacturer and is usually a strictly guarded secret, as the development of a new fiber type can take 10 years and more to develop. Most carbon fiber manufacturers also produce their own precursors or several different ones, depending on the product range. Typical variable influencing parameters in precursor manufacture are the solvents used, types and quantities of co-monomers used, degree of precursor yarn shrinkage, and yarn stretching in the process.

A precursor should be as economic and ecologically friendly as possible, since on the way to carbon fiber about 50% of the precursor's initial mass is lost as gaseous by-products as a result of the chemical conversion processes. An ideal precursor for carbon fibers already has a high carbon content and can be spun into a thin precursor yarn, mechanically stretched, and then converted into a carbon fiber. The fineness of the yarn is further increased by the mechanical stretching, i.e. the fiber diameter is reduced. The stretching also ensures an advantageous alignment of the molecular chains along the fiber axis, which is important for the mechanical properties of the later carbon fiber.

There are two main processes for the production of a precursor filament: *wet spinning* and *melt spinning*. Wet spinning is the most commonly used commercial process for the production of PAN precursors [3-3][3-26]. This is due to the fact that the decomposition temperature of PAN is below its melting temperature. Although the melt spinning process was also developed for PAN, it has not yet been implemented on an industrial scale [3-4].

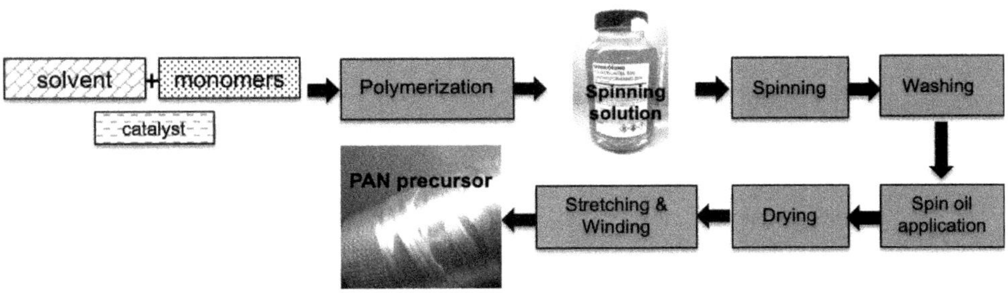

Figure 3.2 Process for precursor production

3.1.1 Wet Spinning

Figure 3.3 shows the typical process flow of a continuous solvent polymerization with subsequent wet spinning. This comprises the two essential steps of the so-called dope preparation (polymerization to polyacrylonitrile) and coagulation (spinning of the PAN). Dope is a general description for a polymer solution is an organic solvent ready for the spinning process (coagulation).

Dope Preparation

To produce dope, powdered acrylonitrile is reacted with other co-monomers, such as methyl acrylate and methyl methacrylate, using a catalyst dissolved in solvents such as dimethyl formamide (DMF) or dimethyl sulfoxide (DMSO) to form polyacrylonitrile. Instead of a solvent-supported reaction, it is also possible to convert the starting materials into a suspension. In order to prevent solidification of the mixture as a result of the polymerization reaction, the temperature of the polymerization reaction must be between 25 and 120 °C. The dope concentration is about 15 to 20% and is pumped directly into the spinning bath (coagulation) [3-5].

Coagulation

In the coagulation stage, the PAN dissolved in the organic solvent as a carrier is pressed through fine nozzles into a chemical spinning bath, where the polymer coagulates and solidifies into thin filaments. Depending on the design, the spinning bath has between 50,000 and 500,000 nozzle holes (spinnerets) with a diameter of 0.05 to 0.25 mm and is operated at line speeds of approximately 3 to 16 m/min [3-5]. The spinning process as such is essentially responsible for the formation of the inner molecular structure of the fiber.

The fiber diameter obtained from the spinning process is usually still too large for use as a carbon fiber precursor and the filaments can also be porous inside. Therefore, the fibers are mechanically stretched and washed several times in the further course in order to remove solvent residues and inhomogeneities, and to achieve the desired fiber diameter. The continuous stretching ensures an advantageous alignment of the molecular chains along the fiber axis, as a basis for the formation of the graphitic molecular structure of carbon during the carbonization process. Finally, a spinning oil is often applied for better textile processing, and the fiber is wound up.

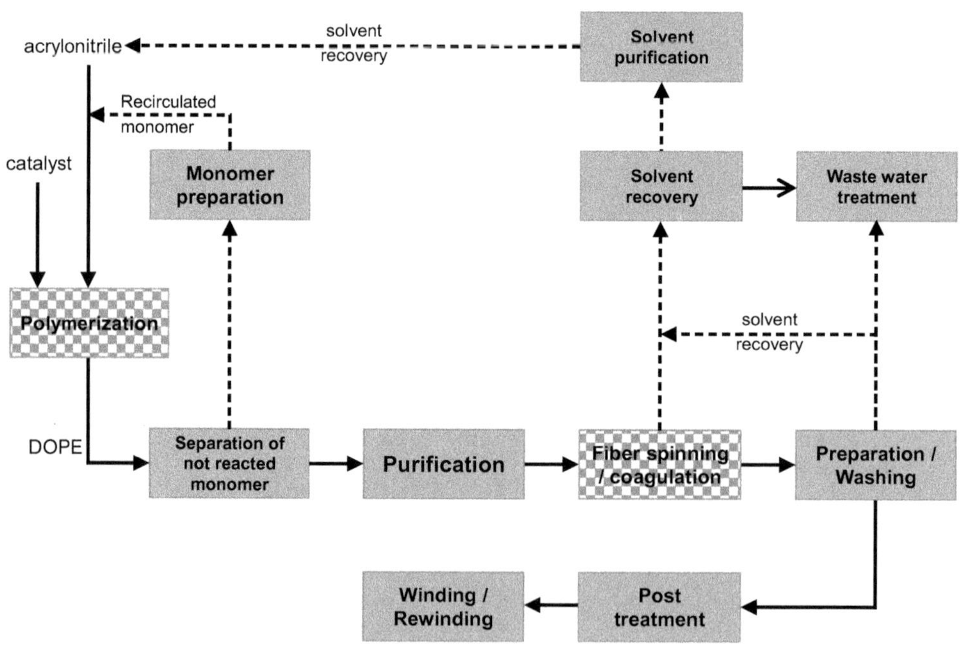

Figure 3.3 Schematic sequence of wet spinning [3-7]

3.1.2 Melt Spinning

Figure 3.4 shows the typical process flow of melt spinning. The process is particularly suitable for precursors such as lignin or polyethylene, whose melting point – unlike with PAN – is below the decomposition temperature. The melting temperature of PAN is around 350 °C. In order to be able to spin PAN in the melt process, PAN is processed with chemical additives and water to form a homogeneous melt, which is extruded into a hot steam pressurized solidification zone through spinnerets (⌀ 60 to 160 µm) at 30 to 70 bar. This process control prevents thermal degradation of the PAN, as the melting point can be lowered to approximately 135 to 185 °C.

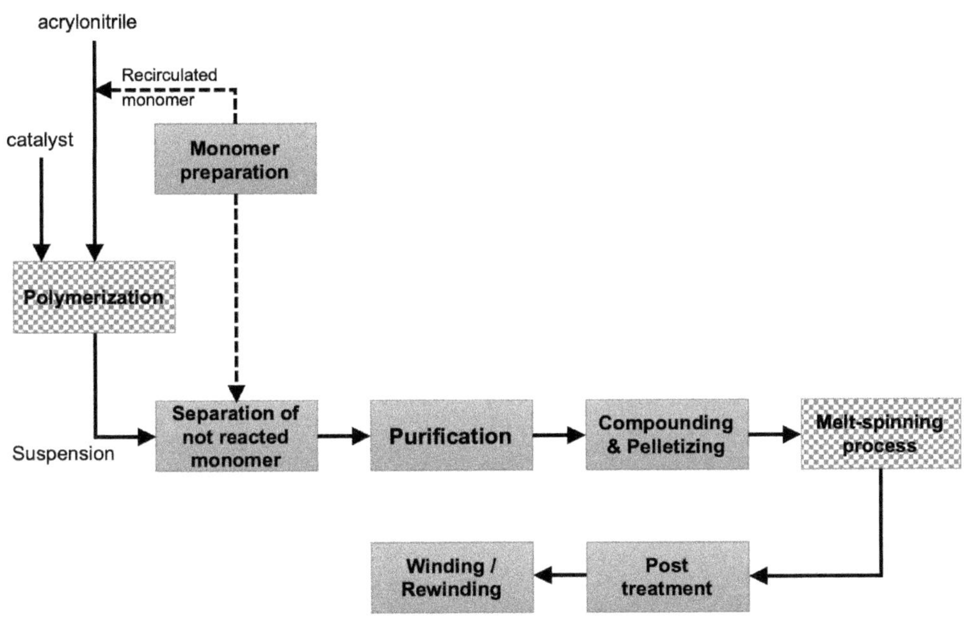

Figure 3.4 Schematic sequence of melt spinning [3-7]

Despite the given processability of PAN in the above-mentioned process, these precursor fibers have more defects inside and on the surface than those produced in the solvent process, so that the fiber properties can be reduced or fluctuate. This is the main reason that the solvent process is preferred for the production of precursors for high-performance fibers [3-8].

A comparison of the two technologies, wet and melt spinning, in Figure 3.3 and Figure 3.4 clearly shows the significantly greater complexity of the wet process. In particular, the post-treatment of wastewater, solvent separation (evaporation process), and solvent recovery are energy- and thus cost-intensive processes. Therefore, the melt spinning process is preferably used for the production of "low-cost" carbon fibers. In addition to the process used, the precursor quality, and the yarn count, the number of filaments (K number) in the yarn bundle (roving/tow) also has a lasting influence on the price of the end product.

While low K numbers (3 to 24 K) are generally used for aerospace applications, fibers from 50 to 480 K are used for wind power and industrial applications. Depending on requirements, the automotive sector uses fibers with both small and high K numbers. A uniform, round cross-section of the PAN precursor is essential for the properties of the carbon fiber, especially for later use in the aerospace industry. Compared to textile PAN, which is used in the clothing industry, this represents a particular challenge in the spinning process and requires a high degree of technical know-how.

3.1.3 Precursors

In the following, different precursors and their manufacturing methods are presented. Basic sources of raw materials (precursors) for the production of carbon fibers are presented in Table 3.1.

Table 3.1 Raw Material Sources (Precursors) for the Production of Carbon Fibers

Resource	C in %	H in %	N in %	O in %	S in %	Yield CF in %
rayon	45	6	–	49	–	20
PAN	68	6	24	–	–	45
pitch (MPP)	94	4	1	0.6	0.4	85
polyethylenes	86	14	–	–	–	75
lignin	62	6	–	32	–	40

3.1.3.1 Polyacrylonitrile (PAN)

Polyacrylonitrile (PAN) is a thermoplastic, semi-crystalline plastic with a planar chemical structure (Figure 3.6). The starting material for PAN is acrylonitrile, which is produced by the Sohio process, a catalytically controlled conversion of propene with ammonia and pure oxygen (Figure 3.5) [3-21]. The reaction is also called ammoxidation of propene. Acrylonitrile, with acetonitrile and hydrogen cyanide (prussic acid) as by-products, is formed by splitting off water, which must be separated.

$$CH_2{=}CH{-}CH_3 + NH_3 + 1.5\ O_2 \longrightarrow CH_2{=}CH{-}CN + 3\ H_2O,\ \Delta_R = -\ 516\ \text{kJ/mol}$$

Figure 3.5 General reaction equation for the production of acrylonitrile

Despite the thermoplastic character of PAN, it does not melt but decomposes before the melting range from 300 °C is reached. The melting range can only be achieved without prior decomposition by extreme heating rates of ≥ 50 °C/min. Almost all PAN precursors are copolymers of acrylonitrile with other monomers such as methyl acrylate and methyl methacrylate. Three essential properties make PAN an ideal precursor in carbon fiber production:

- PAN has a continuous carbon chain.
- The chemical structure is optimal for ring formation (cyclization reaction).
- The carbon content in acrylonitrile (CH_2=CH-CN) is already about 68%.

The high proportion of carbon is important, since about 50% of the molecular mass is lost during the subsequent processes.

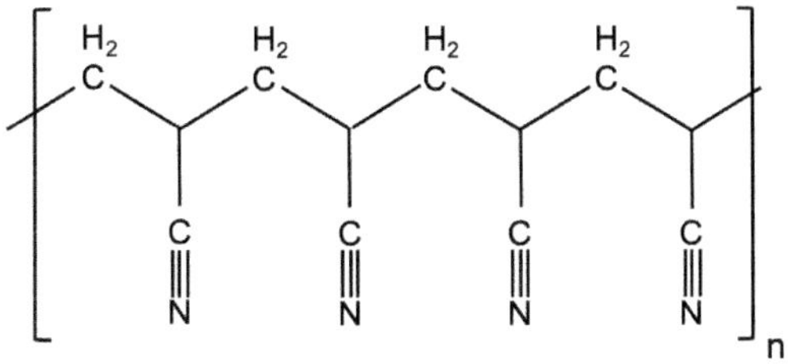

Figure 3.6
Chemical structure of polyacrylonitrile (PAN) [3-9]

The cyano (CN) group in the molecule has a distinct dipole moment due to the differences in the electronegativity of the nitrogen atom (N) and the carbon atom (C). This dipole moment leads to an electromagnetic interaction between the CN groups of the long-chain PAN molecules, which accounts for the chemical structure of the PAN fiber and its special properties (strength, etc.) (Figure 3.7).

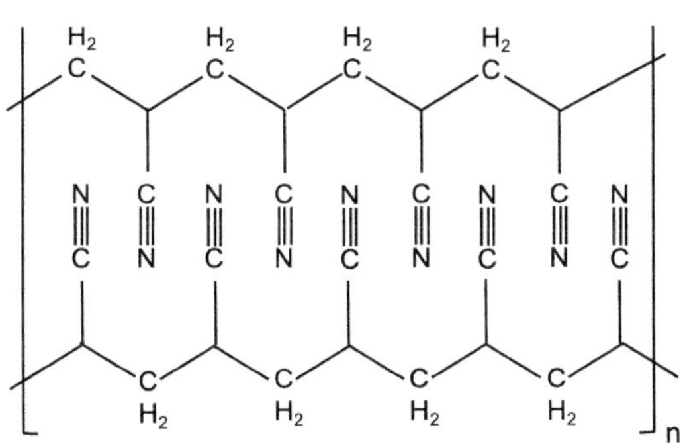

Figure 3.7
Chemical structure of adjacent PAN fibers [3-10]

In this context, it should be pointed out once again that the PAN precursor for carbon fibers differs chemically from textile polyacrylonitrile, which is produced for the manufacture of clothing.

3.1.3.2 Mesophase Pitch

Compared to PAN-based carbon fibers, mesophase pitch carbon fibers have a higher modulus of elasticity and lower tensile strengths. During the manufacturing process, liquid crystalline polyaromatic hydrocarbons are aligned in the direction of the fibers during the spinning process. These polyaromatic hydrocarbons are produced by heat treatment of fusible pitches. This heat treatment causes condensation to build up polyaromatic hydrocarbons with a molecular weight between 1000 and 2000. At a temperature of about 400 °C, these polyaromatic hydrocarbons have the ability to form spherulitic liquid crystals of the carbon fiber. The surface tension of these spherulites causes touching liquid crystals to coalesce, i. e. combine to form a larger liquid crystal. During the subsequent spinning process, fibers with a high degree of orientation in the axial direction are produced. Analogous to the PAN production path, oxidation makes the fiber infusible. Only then can it be thermally degraded into a carbon fiber in the subsequent carbonization process.

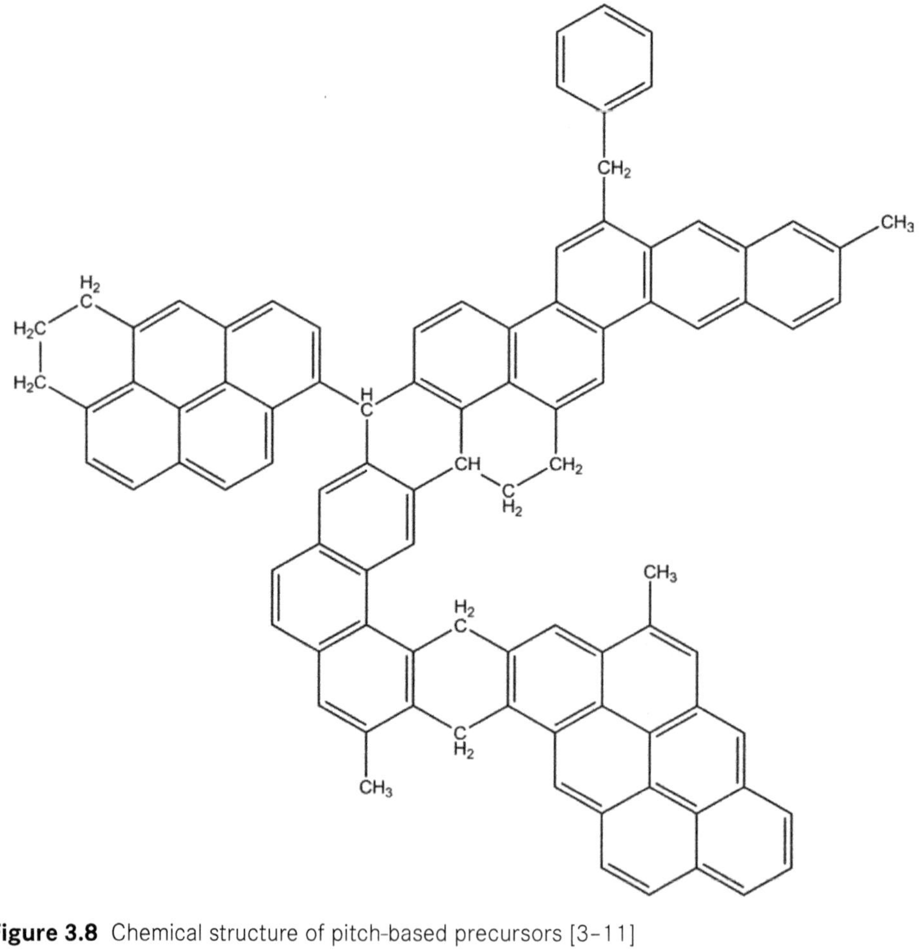

Figure 3.8 Chemical structure of pitch-based precursors [3-11]

For high performance fibers, many filaments are usually combined into bundles, so-called roving or tow, of 1000 to 400,000 individual filaments and wound into bobbins weighing 50 kg. These coils are then fed to the next step, carbonization.

3.1.3.3 Further Precursors

The ideal precursor for a carbon fiber should have as high a carbon content as possible, since on the way to the carbon fiber about 50% of the precursor's initial mass is lost as gaseous by-products as a result of the chemical conversion processes. The precursor must be easily spun into a fiber and convertible into a carbon fiber. In addition, this precursor should be as economic and ecologically friendly as possible.

In addition to the classical PAN- and pitch-based precursors, there are some promising new approaches for suitable starting materials on other chemical bases, such as polyethylene, cellulose, and lignin. These approaches are explained in more detail in Chapter 7.

3.2 Production of Carbon Fibers

After the precursor has been manufactured, it is converted into carbon fiber under further temperature influence in three essential steps:

1. *Oxidative stabilization* (200 to 300 °C – formation of a stable basic structure, which is a prerequisite for processability at high temperatures)
2. *Carbonization* (< 1600 °C – removal of all foreign atoms and formation of turbostratic structure from carbon atoms)
3. *Graphitization* (> 2000 °C – improvement of the orientation of the carbon atoms in the fiber leads to higher properties of the carbon fiber)

Graphitization is an optional step in the production of certain carbon fiber types. After the carbonization/graphitization, a surface treatment of the fiber follows.

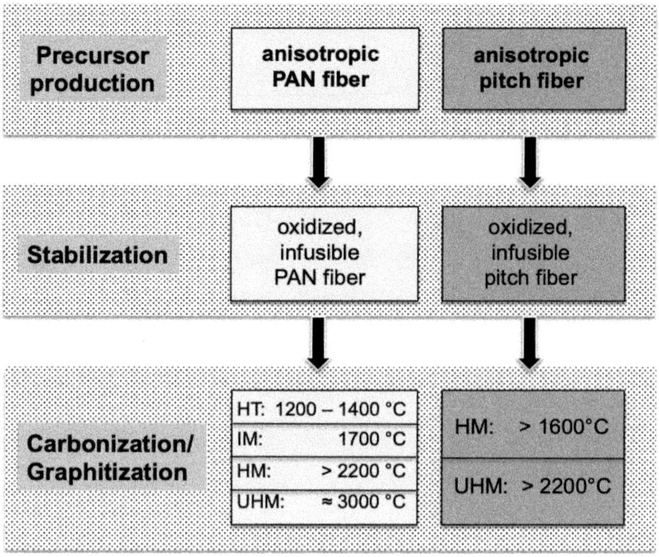

Figure 3.9 Production of carbon fibers

The complete process flow for carbon fiber production is shown schematically in Figure 3.10.

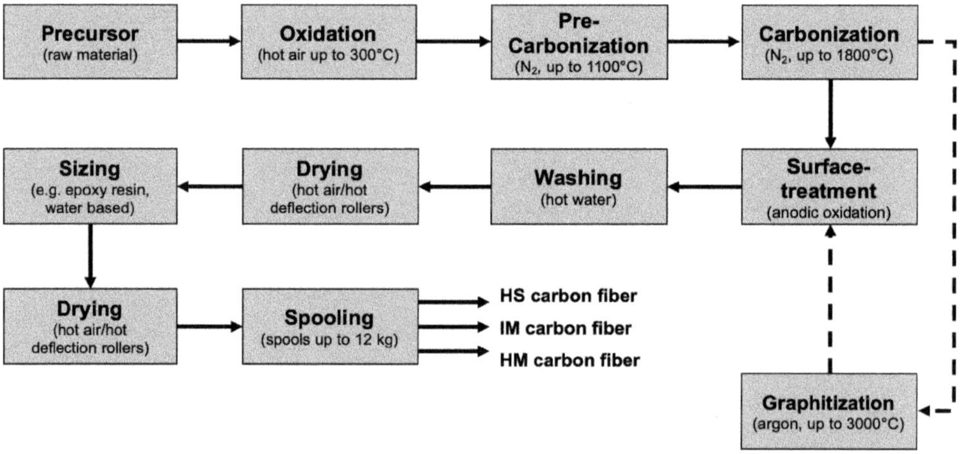

Figure 3.10 Schematic sequence of the entire carbon fiber manufacturing process

Important manufacturers of such process technology, which includes not only furnace technology but also the entire exhaust gas treatment and heat recovery system, include *Eisenmann Thermal Solutions GmbH* and *Harper International Inc.* Further suppliers of process technologies and precursors are listed in Table 3.10 and Table 3.12, in Section 3.5.1 and Section 3.5.4, respectively.

3.2.1 Precursor Stabilization/Stabilization Process

Before the thermoplastic PAN precursor can be converted into carbon fiber at high temperatures, it must first be thermally stabilized under oxidizing conditions and moderate heat (200 to 300 °C). The schematic sequence is shown in Figure 3.11. In Figure 3.12 and Figure 3.13 reactions that take place in principle are shown. In this stabilization stage, which lasts approximately 30 to 120 minutes, the molecule is first cyclized and dehydrogenated. Control of this reaction is critical, because it is strongly exothermic and requires an exact temperature and air guidance within the oxidation furnaces. This is decisive for the later quality of the carbon fiber.

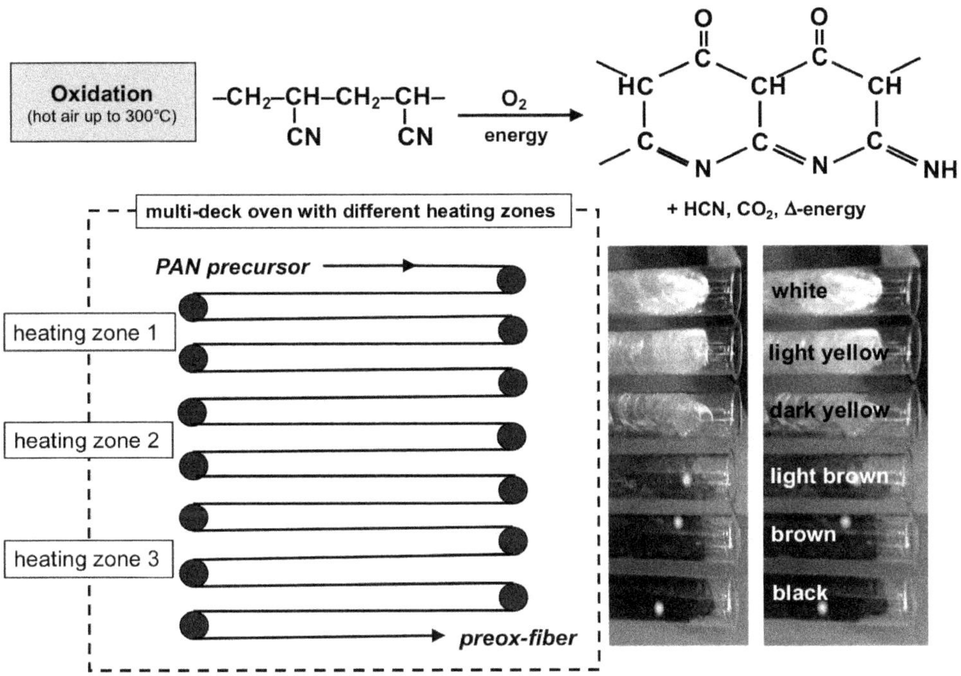

Figure 3.11 Schematic sequence of precursor stabilization

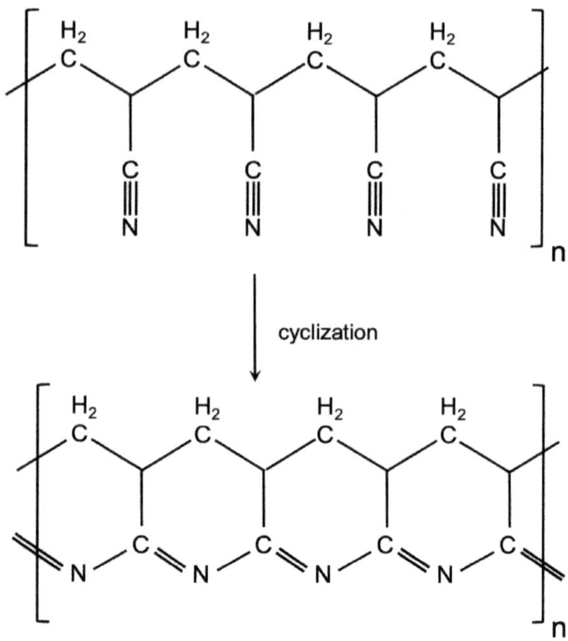

Figure 3.12 Reactions during thermal cyclization of PAN [3–12]

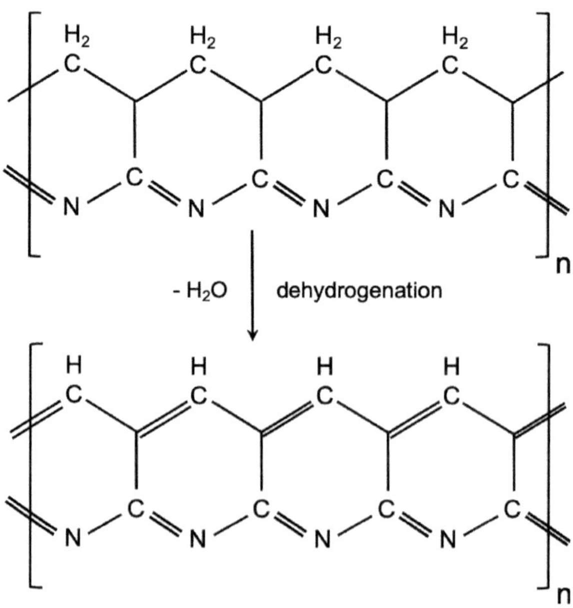

Figure 3.13 Dehydrogenation of PAN [3–13]

The absorption of atmospheric oxygen and the elimination of water cause a further chemical reorientation of the molecule to a so-called conductor polymer, from which pyridine six-rings are formed by oxidative cyclization (Figure 3.14). At the same time, the fiber is further stretched during the process to optimize molecular and structural alignment.

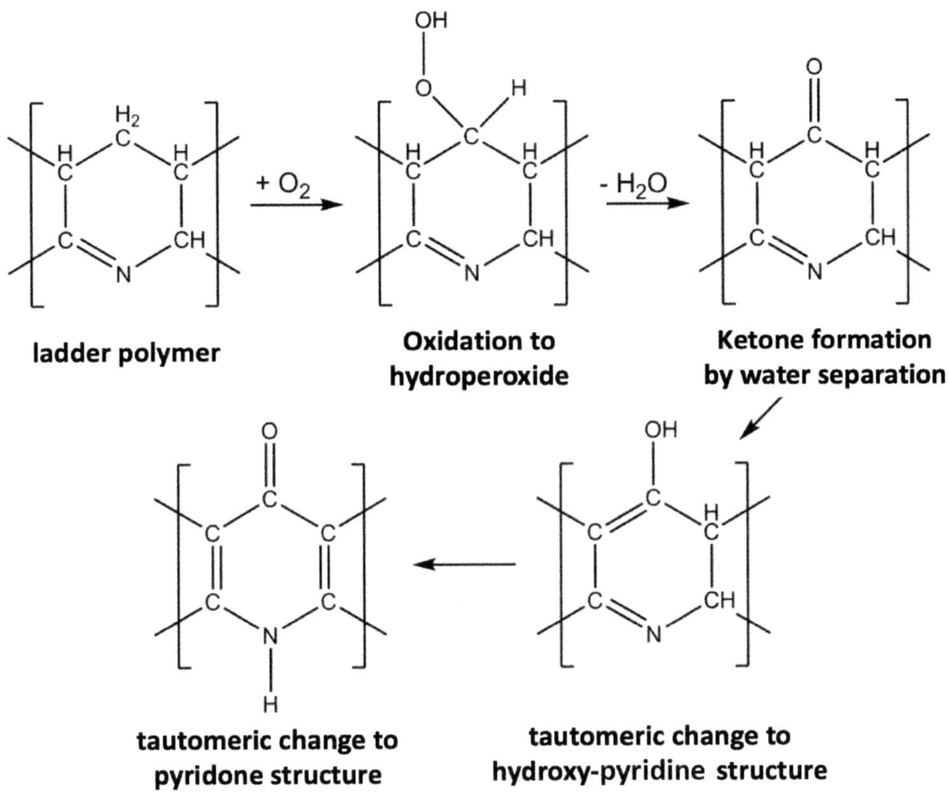

Figure 3.14 Oxidative stabilization of polyacrylonitrile (PAN) [3-14]

The stabilizing chemical reactions are very complex, include several steps, and sometimes even take place simultaneously. The resulting reaction heat must be controlled in such a way as to prevent overheating and possible destruction of the fiber. A wide variety of systems and technologies are used commercially. In some processes the fibers are fed through several heating chambers connected in series, in others via heated rollers or through heating baths. In addition, further gases can be added to the surrounding hot air to accelerate the chemical process. During the process, the color of the PAN precursor changes from white through several levels of brown/red-brown to black (Figure 3.15).

Figure 3.15 Color change of polyacrylonitrile during stabilization

The PREOX fiber obtained is now infusible, which is the prerequisite for the subsequent carbonization process.

This textile fiber is characterized by a carbon content of around 60% and already has a number of interesting properties:

- Excellent chemical resistance and very good electrical insulation.
- Non-combustible, oxygen index LOI 50 to 60 (LOI = Limiting Oxygen Index), significantly higher than that of other organic flame-retardant fibers.
- Not meltable.

The resulting PREOX fibers can either be processed directly into carbon fibers in the process or used as continuous fibers or as crimped staple fibers. PREOX fiber types widely used in the market are PYROMEX™ fiber (Tejin Carbon), PANOX™ fiber (SGL Carbon), and Zoltek OX™ fiber (Zoltek).

3.2.2 Carbonization and Graphitization

The thermally stabilized PREOX fiber is now heated for several minutes in a continuous furnace under inert gas atmosphere (oxygen exclusion) to temperatures of 1000 to 3000 °C (1830 to 5500 °F). The exclusion of oxygen prevents the fibers from burning under the extreme temperatures. In addition, depending on the design, the furnace is operated under a slight overpressure of protective gas to prevent oxygen from penetrating the open ends of the furnace.

During the carbonization process, the pyridine chains previously obtained combine to form molecular bands (graphitic structures) by splitting off hydrogen cyanide and nitrogen. At the same time, an orientation in the third dimension takes place, so that a graphite-like structure develops. Depending on the treatment temperature, a distinction is made between three fiber classes with different properties:

- High-strength fibers (HT/HS type), approx. 1200 to 1500 °C.

- Intermediate modulus fibers (IM type), approx. 1500 to 1800 °C.
- High modulus fibers (HM, UHM types), up to approx. 3000 °C.

In some processes, two furnaces connected in series with different temperatures are also used in order to better control the heating rate and the temperatures acting on the fibers during carbonization (Figure 3.16).

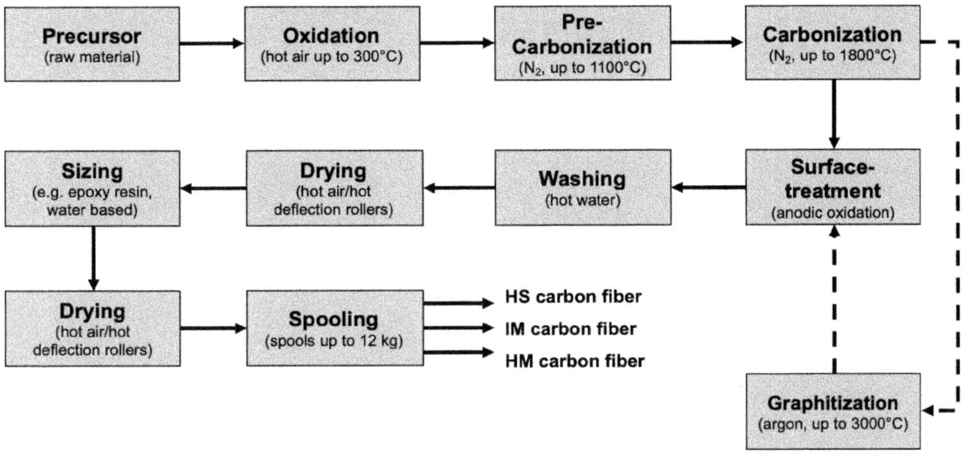

Figure 3.16 Schematic process of carbonization

The graphitic ribbon structure increases with increasing treatment temperature in the carbonization process (see above). This alignment of the ribbons along the longitudinal axis of the fibers is the prerequisite for the exploitation of the covalent and extraordinarily stable carbon-carbon bonds (C–C bonds) in order to achieve high fiber strength and a high modulus of elasticity. Nevertheless, these ribbons must not run endlessly parallel to the fiber axis and have a perfect graphite structure. This would have the consequence that the layers of the graphite structure (see Chapter 2) would slide off of each other and the high bond energy of the covalent C–C bonds could not be exploited, i.e. the defects in the lattice structure that occur primarily on the fiber surface are essential for the property profile of carbon fibers. At these defects the carbon atoms exhibit free valences and thus represent areas of increased energy and reactivity. With increasing treatment temperature in the manufacturing process, the graphite structure of the fiber becomes more perfect, as both the orientation of the ribbons in the longitudinal direction is increased and defects "heal". This reduces the coupling possibilities for matrix polymers (thermoplastics or thermosets) for the production of a fiber composite material. As a result, the *fiber matrix adhesion* (*FMA*), which is essential for composite properties, decreases.

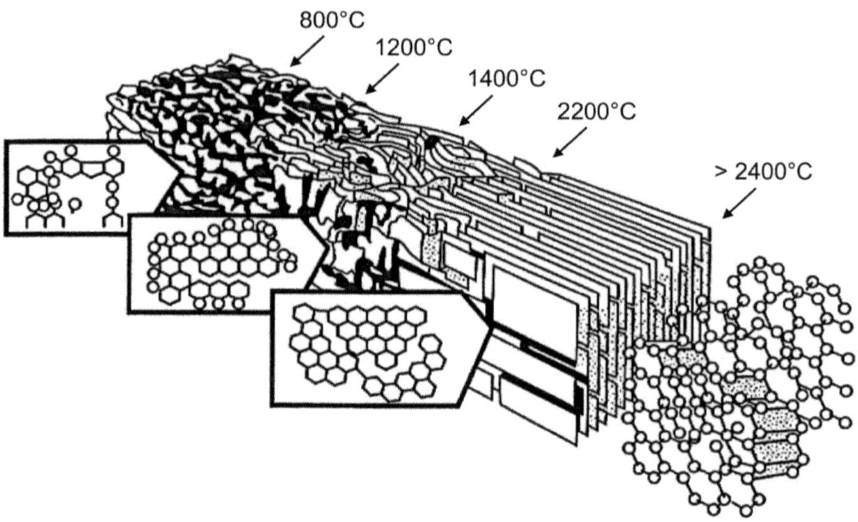

Figure 3.17 Change in the chemical structure of PAN under the influence of heat [3-15]

Nevertheless, even the not completely saturated carbon atoms on the surface are chemically extremely inert, so that a coupling of the matrix polymer is only possible via the significantly weaker van der Waals forces, thus leading to a lower mechanical property profile in the fiber composite. In order to improve the fiber matrix coupling and thus the properties of carbon fiber composites, numerous methods for surface treatment (modification) have been developed.

■ 3.3 Surface Treatment and Fiber Sizing

The surface of the carbon fiber must be activated after the carbonization process to allow sufficient bonding of the matrix polymer and to improve the fiber-matrix adhesion, which is so important for the mechanical properties of the composite material. Ideally, chemical bonds are formed between the active surface groups formed in the oxidative treatment and the reactive groups of the matrix resin. The different surface activation processes can be divided into *non-oxidative* and *oxidative* methods. The oxidative methods are mainly used in industrial processes.

Non-Oxidative Processes

Important modifications of the carbon fiber using non-oxidative processes are the following:

- Whisker formation (enlargement of the surface)

- Electropolymerization (targeted modification of the layer between fiber and matrix, which has a major influence on the mechanical properties of the composite, e.g. elastification)
- Coating of the fiber with pyrocarbon (increase of the mechanical serration; some methods lead to the formation of VGCF, i.e. vapor-grown carbon fibers)

Oxidative Processes

Oxidative methods are the most common methods today for the targeted chemical surface treatment of carbon fibers. Functional chemical groups (e.g. carbonyl groups) are formed which can be coupled to matrix polymers such as epoxy resins or thermoplastics, resulting in improved fiber-matrix adhesion.

It is distinguished into the following:

- Wet oxidation with various dissolved or liquid oxidizing agents, e.g. nitric acid, sodium hypochlorite solution
- thermal oxidation
- anodic oxidation

The most suitable commercial technical processes are anodic and thermal oxidation.

The thermal oxidation of the carbon fiber takes place in gaseous media such as oxygen, oxygen/nitrogen mixtures, carbon dioxide, or others at temperatures of 400 to 1100 °C. The treatment in gaseous media makes subsequent washing processes necessary for anodic oxidation superfluous. Depending on temperature and gas composition, the residence time ranges from a few seconds to minutes.

In anodic oxidation, the fiber is passed through an electrolyte bath and switched as an anode. Diluted acids, bases, or salt solutions (e.g. ammonium sulfate $(NH_4)_2SO_4$) are used as electrolytes. The anode potential is typically between 1 and 12 volts with dwell times between 0.5 and 5 min. A schematic sequence is shown in Figure 3.18. Following the surface treatment, excess electrolyte is removed from the fiber surface in a water bath, as this would have a disturbing effect in the further processing of the fiber [3-17].

In addition to the formation of functional groups by the chemical reaction of carbon atoms on the surface, oxidative treatment usually also leads to a change in the actual surface of the carbon fiber. Therefore, the process must be well controlled to avoid damage to the fiber. For example, excessive oxidative attack and degradation can result in notches that reduce the fiber strength.

Due to the processes and technologies used by the various manufacturers, there may be differences in connectivity between fiber types and manufacturers. The manufacturers usually do not specify how the fiber surface was activated.

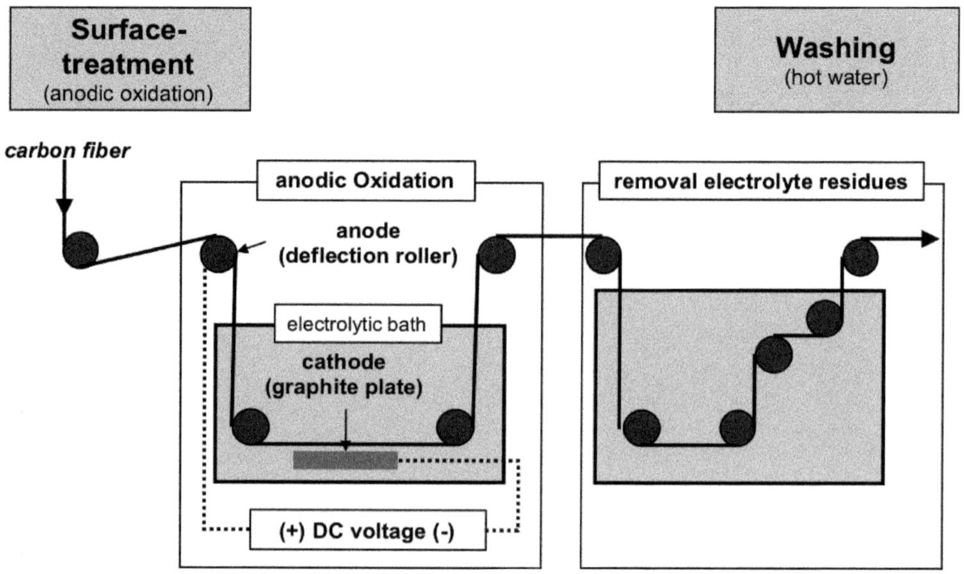

Figure 3.18 Surface treatment by anodic oxidation

3.3.1 Sizing (Preparation)

After surface activation, a *"sizing"* (also called *preparation*) is applied to the carbon fiber. The preparation application has several tasks and can be understood as an auxiliary material for the further processing of the fiber, which usually makes the further textile processing of the carbon fibers possible in the first place:

- Improvement of processability in the manufacture of flat or three-dimensional textile semi-finished products
- Improvement of wettability with the matrix
- Modification of the interface between fiber and matrix to improve the mechanical properties of the composite material (increase of the utilization of the fiber properties in the composite material)

The sizing should not be confused with a binder that has a different function and chemical structure and is only applied at a later point in time if required. The carbon fiber strands are first dried before the sizing application, e.g. using heated deflection rollers (contact drying). This supports an even take-up of the sizing in the next sub step. Afterwards, a new drying takes place, mostly by means of hot air, and the final winding of the finished carbon fiber on cardboard cores with up to 12 kg spool weight. In Figure 3.19 a schematic sequence of the sizing order is shown.

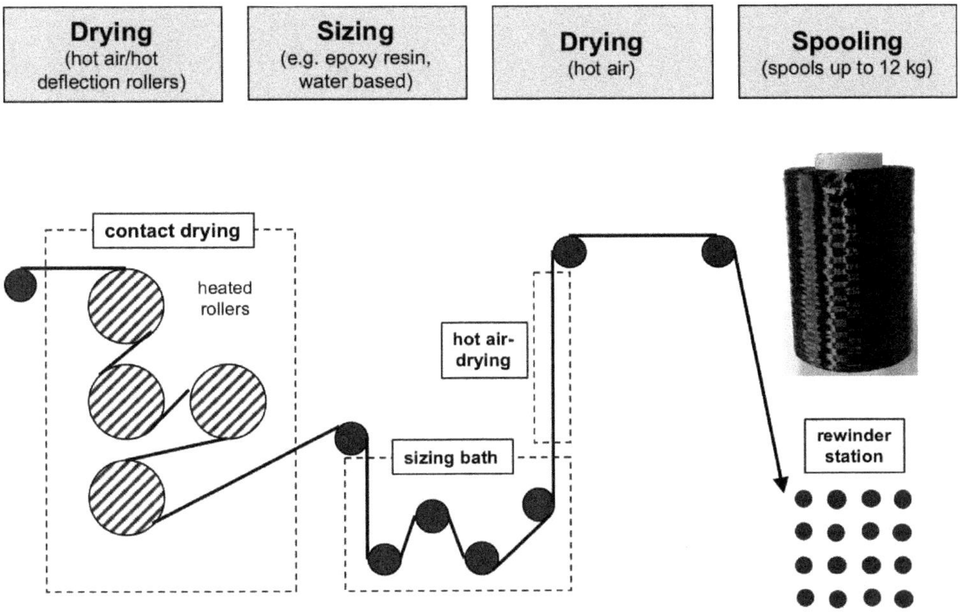

Figure 3.19 Schematic procedure of the sizing order

The different textile processing processes of dry carbon fibers sometimes place very different demands on the yarn and thus also on the preparation. In the production of unidirectional prepregs (UD prepregs), the width of the fiber achievable by spreading is important for achieving the desired fiber basis weight without the occurrence of gaps in the later prepreg (so-called minigaps). Therefore, the sizing must not cause too much adhesion between the filaments, i.e. it must not stick them together. Nevertheless, it must protect fiber bundles from mechanical damage and must not impede the penetration of the matrix resin into the bundles. This technical balancing act is solved on the one hand by adjusting the sizing level, i.e. the amount of preparation applied, and on the other hand by chemically adapted sizings. While in the early days of carbon fiber production, water-soluble preparations (nylon-based) were used that were known from the textile industry, the sizings were adapted over time so that more epoxy resin (EP)- or polyurethane (PU)-based preparations are currently used [3-17].

In textile processing methods such as weaving, braiding, or filament winding, the fiber bundles are subjected to relatively high mechanical stress. In this case, the sizing must provide good thread closure to protect the filaments so that the fiber bundles have low frictional resistance at the thread eyelets and deflection rollers. Possibly broken filaments must neither accumulate nor lead to a break in the fiber strand. Nevertheless, the thread closure must not hinder the subsequent impregnation of the textile semi-finished product with matrix polymer.

In order for a fiber bundle of 1 to 48 K individual filaments to exhibit uniform properties such as friction, spreadability, or thread closure, the outer and inner filaments of the bundle must be evenly coated with the preparation. To solve this technically demanding task, special application methods are used on the one hand, and on the other hand the preparation solutions must be optimized with regard to their impregnation and wetting properties. The processability in different processes, the impregnation capability, and the influence of the sizing on the fiber matrix adhesion must be very well coordinated.

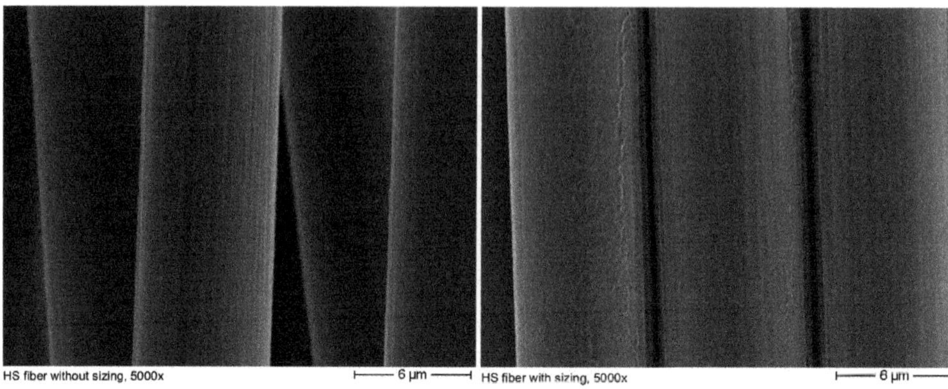

Figure 3.20 SEM image of HS carbon fiber without (left) and with (right) sizing, 5000× magnification

In Figure 3.20 longitudinal grooves can be seen on the filaments, which usually come from spinning the precursor. Ideally, these should not change during the spinning process. Experience shows, however, that the spinnerets can get dirty during production, especially towards the end, and the "grooves" change their appearance. Figure 3.20, left, shows filaments without sizing - only the spinning grooves are visible. On the right of Figure 3.20 wavy lines can be seen on the spinning grooves, which result from coating with sizing. When applying the sizing, it is very important that it is applied absolutely evenly and distributed very finely on each individual filament with the very small quantities. For a yarn with 800 tex and a size level of 1%, 8 g sizing is distributed over a yarn length of 1000 m to 12,000 filaments.

With a bad sizing distribution or only local wetting, the filament bundle would only be wetted around the outside and the tow in the geometry would be fixed on a "shoelace cross section". Such sizing systems, which in the past were often referred to as web sizing, are now hardly in use or no longer used at all.

It is assumed that the following processes play a major role in fiber-matrix connection:

- The matrix couples to the sizing and the sizing couples to the fiber. The sizing on the fiber reacts with the matrix (primary effect).
- The sizing passes into the matrix and this mixture couples to the activated fiber surface.
- The matrix couples directly to the fiber surface.

Which of these processes dominates in each case has not yet been conclusively clarified. However, the interaction between fiber surface, sizing, and matrix, also known as interphase, is decisive for the properties in the composite. It is therefore crucial to ensure fiber quality by means of a suitable specification that includes both the fiber properties and the production parameters. This is the only way to ensure that high-quality carbon fibers can be produced over a long period of time with reproducible properties and that good characteristic values can also be achieved in the composite material.

Interphase is a very thin transition region (approx. 110 to 180 nm, Figure 3.21, [3-25]) between fiber and matrix. Although the thickness of this intermediate layer between the carbon fiber and the matrix system is very small, it plays a decisive role in the load transfer from the matrix to the fiber and influences design criteria such as strength, stiffness, and weight. Good interface properties are essential to ensure efficient load transfer from the matrix to the fibers, reduce local load concentrations, and improve overall mechanical properties. The properties of the interphase are decisively determined by good interfacial adhesion between the fiber and the matrix. This interfacial adhesion results from chemical and physical interactions, which strongly depend on the molecular interfacial structures and interfacial interactions. With high activity of carbon fiber surface and matrix, these tend to form stable chemical compounds, which leads to a strong adhesion between the binding partners. The relatively larger surface area of the carbon fibers compared to the matrix increases the impregnation and wetting of the fibers as well as the mechanical interaction between fiber and matrix, which leads to a further improvement of the surface adhesion. Although a high stiffness of the interphase is very important for an effective load transfer between fiber and matrix, a brittle behavior of the interphase, which does not represent a barrier for crack propagation in the composite, can lead to a catastrophic failure of the laminate. In contrast, a ductile interphase can prevent crack propagation between the fibers and thus protect the laminate from premature failure. A balance between stiffness and ductility of the interphase is therefore crucial for the design of composites and their desired mechanical properties.

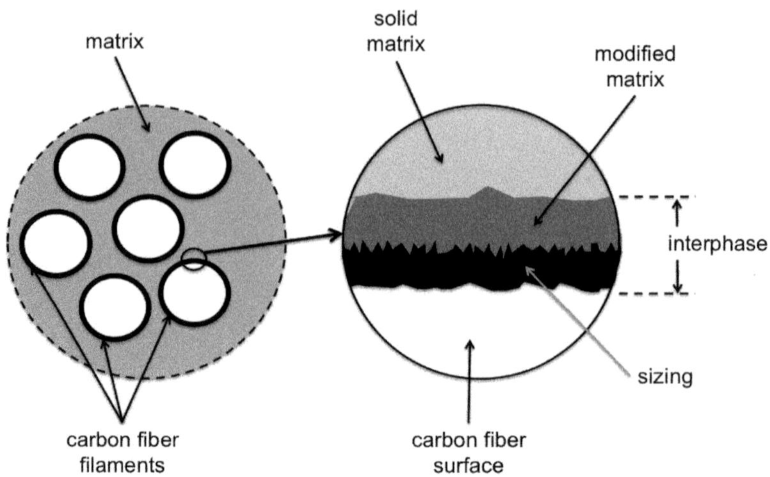

Figure 3.21 Schematic diagram of the interphase (after [3-25])

3.3.2 Sizing Levels and Sizing Types

Another factor influencing fiber matrix adhesion (FMA) is sizing level, i.e. the amount of preparation applied to the fiber surface. This is between 0.8 and 2% for carbon fibers, depending on fiber type, manufacturer, and application.

Table 3.2 shows examples for sizing, application, sizing level, and their compatibility with different matrix polymers. In particular, the chemical compatibility of the sizing with the matrix used is essential for good fiber-matrix adhesion. If these two components do not match chemically, the matrix polymer is almost not bound to the fibers and the composite material fails prematurely. An example of this unpleasant effect could be observed in 2010 on a Norwegian motor catamaran, where the water-side lower deck began to delaminate considerably due to the above-mentioned incompatibility and wave impact during operation. Therefore, new sizing or matrix systems (thermoplastics and thermosets) are often tested for compatibility nowadays (see Chapter 4).

Table 3.2 Examples for C-Fiber Sizings [Source: Hexcel Corporation]

Designation	Size Compatibility	Recommended Use	Sizing Level
Unsized (without sizing)	epoxy, phenolic, polycarbonate, polyurethane, polyester, polysulfone, cyanate ester, vinyl ester, nylon, BMI, PES, PEEK, PEKK, PES, PVC, polyimide, polypropylene	prepreg UD tape	0

Designation	Size Compatibility	Recommended Use	Sizing Level
G	Epoxy, phenolic, polyurethane	weaving, prepreg UD tape	0.8–1.2 0.2–0.4
GP	Epoxy, phenolic, vinyl ester, polyurethane, cyanate ester, BMI	weaving & filament winding, prepreg UD tape	0.8–1.2 0.2–0.4
H	epoxy	weaving	0.8–1.2
R	epoxy, polyester	filament winding	1.2–1.6
GS	epoxy, vinyl ester, polyurethane	prepreg UD tape	0.3–0.7
E	epoxy, polyester	prepreg UD tape, weaving, pultrusion, filament winding	0.9–1.3
J	epoxy	weaving	0.8–1.2

Usually three different groups of sizing types are used for carbon fibers:

a) Non-Permanent Processing Sizing

A non-permanent processing sizing is a sizing that is applied to a yarn in order to be able to carry out a textile process afterwards. After processing, the sizing is washed out again. There are areas such as thermoplastic applications and BMI resin systems where it can be advantageous not to have sizing on the carbon fiber. These sizing types therefore have no influence on the fiber matrix connection. However, such a sizing type is not in high demand on the market and is only offered to a limited extent for carbon fibers.

b) Permanent Processing Sizing

Permanent processing sizing means sizing that is permanently applied to the fiber and remains there. It must therefore be compatible with the matrix used later. Sizings are often used on the basis of epoxy, polyurethane, phenol, PVA, or PVE. Information on the compatibility of the various sizings with matrix polymers is provided by the fiber manufacturers. Sizings based on epoxides or polyurethanes are the most widely used today. These sizing types usually harmonize with the commonly used epoxy resin systems over all temperature ranges (> 190 °C). However, there are also special sizings for thermoplastic matrix systems, where decomposition of the sizings must be prevented due to the higher processing temperatures, so that the resulting decomposition products do not negatively influence the fiber-matrix bond. Permanent processing sizings are the most frequently used sizings for carbon fibers.

c) Functional Sizing

Functional sizing refers to sizing types, such as binder sizing and high-temperature sizing, which open up additional processing properties or additional applications for the carbon fiber. These sizing types can also be applied additionally on top of an already applied sizing or directly to the fiber surface. A typical representative is, for example, the additional sizing which is applied to incorporate carbon fibers into rubber (synthetic rubber), e.g. for timing belts. Another example would be the application of a binder that allows preforms to be fixed accordingly or individual yarns to be deposited and fixed.

It can be assumed that new types of sizing will be developed for new applications of carbon fibers or due to changes in storage and processing processes. While some sizings can only guarantee constant properties for up to one year from the date of manufacture, other preparations offer up to five years. Basically, carbon fibers should be stored under moderate climatic conditions (< 35 °C, < 50% RH) and under exclusion of direct sunlight or moisture in order not to change the properties of the sizing. Changes may not only affect fiber-matrix adhesion and composite properties, but also the processing behavior of the fiber, e.g. spreadability in textile processes and impregnability with matrix polymers [3-18].

The following inherent properties of carbon fibers without sizing have unlimited validity:

- Tensile strength
- Tensile modulus (Young's modulus)
- Tensile elongation
- Fiber density
- Weight per unit of length (mass per unit length)

3.4 Properties and Types

In the following, the chemical structure and typical properties of carbon fibers in general are explained. The different fiber types and their specific properties are presented in comparison with each other as well as with other materials. The chapter concludes with a list of the manufacturers and production capacities available on the market.

3.4.1 Properties of Carbon Fibers

Like graphite, carbon fibers are layered materials due to their structure with two-dimensional covalent bonds. The basic structural element is the graphite layer or graphite single crystal (Figure 3.22). The individual layers, as can be seen in Figure 3.22, are slightly shifted against each other and are held together by *van der Waals* forces. In principle, a layered structure with an ordered ABA sequence is formed. These layers consisting of carbon rings, which are formed during oxidative stabilization and subsequent carbonization, and their high degree of orientation in the axial direction decisively determines the strength and modulus of the fiber in the longitudinal axis. The theoretical values for strength and modulus can be calculated from the bond energies. In the fiber direction the theoretical modulus $E_{||}$ is 1050 GPa. Crosswise to the fiber axis, a much smaller modulus $E_{\perp}$ is achieved (due to the lower binding energies) which amounts to only 37.5 GPa [3-17]. The prerequisite would be that there are no defects in the lattice structure of the fiber, which does not occur in reality, so that there is no completely ordered stacking of the layers ABA.

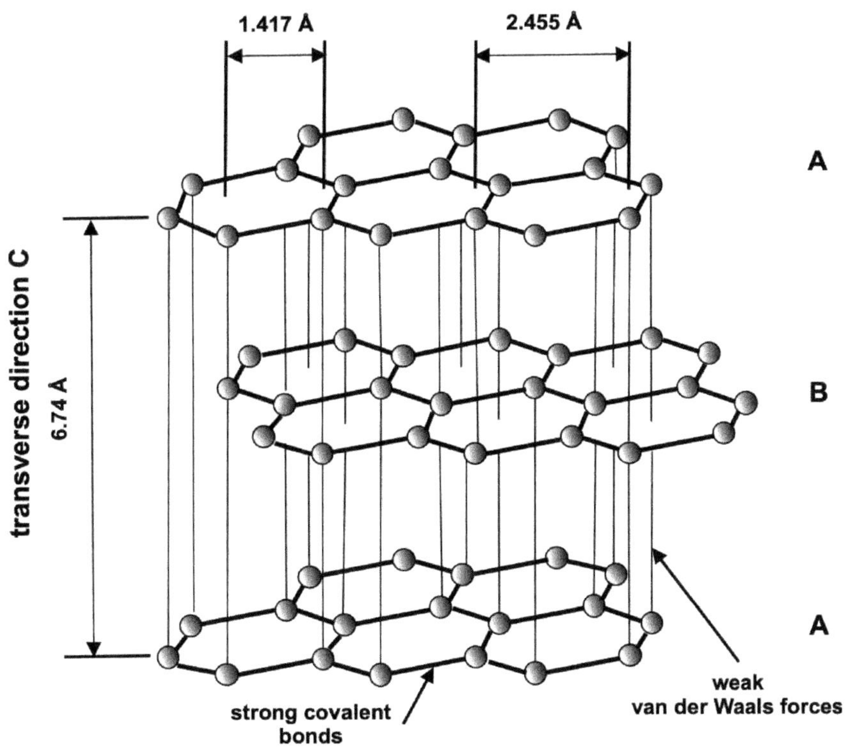

Figure 3.22 Graphite single crystal schematic (according to [3-17])

The very different behavior of carbon fibers as a function of the spatial direction can also be seen in other characteristic values, such as electrical conductivity and the thermal expansion coefficient α. Due to the layer structure, electrons can only move along the graphite layers, i. e. in the fiber direction. Depending on the fiber type, very low electrical resistances can be found here, while the electrical resistance rises sharply across the fiber axis. The electrical resistance depends essentially on how well the graphitic structure is formed in the fiber strand. Pitch-based fibers are particularly suitable here. The same applies to the coefficient of expansion. Along the fiber axis, carbon fibers show a negative coefficient of expansion, i. e. they contract with temperature increase, while they expand transversely to it.

Table 3.3 Thermal Expansion Coefficient α at 20 °C and Electrical Conductivity [3-16] [3-17] [3-21] [3-22] [3-23]

Property	Unit	HT Fiber	HM Fiber	Aluminum	Steel	Titanium	Copper	Wood
$\alpha_{\parallel}$	10^{-6}/K	−1.0	−1.5	23.2	13	10.8	16.5	8
$\alpha_{\perp}$		10	15					n. a.
electrical conductivity$_{\parallel}$	S/m	(0.25–10)·10^3		$3.6 \cdot 10^7$	$5 \cdot 10^6$	$2.5 \cdot 10^6$	$5.8 \cdot 10^7$	$1 \cdot 10^{-9}$ to $1 \cdot 10^{-4}$
electrical conductivity$_{\perp}$		0.5–10						

Carbon fibers are 100% paracrystalline. Like other crystalline substances (e. g. sodium chloride NaCl or gallium arsenide GaAs), graphite crystals form a crystal lattice. While the lattice of NaCl is perfectly regular from any point of the crystal, paracrystalline materials such as carbon fibers behave only approximately so. In principle, the individual components in the lattice are arranged regularly, but the exact appearance of the rest of the crystal lattice cannot be deduced by looking at an elementary cell, i. e. the smaller the area under consideration, the more regular the arrangement of the lattice appears.

Carbon fiber	Structure	Properties
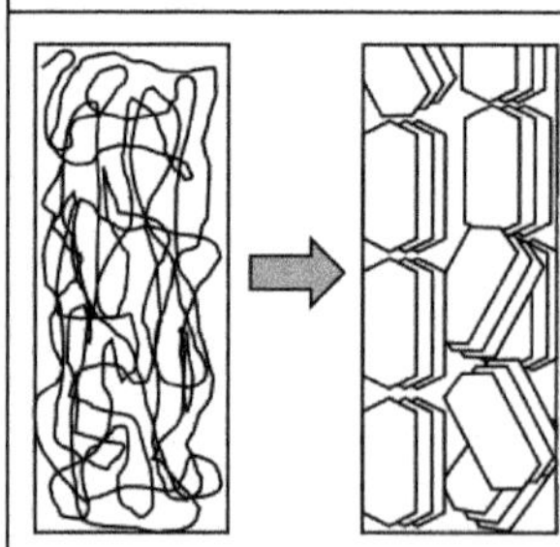 Stabilization / Carbonization	▪ 2D covalent bonds ▪ 100% paracrystalline ▪ high orientation of layers in axial direction	▪ high tensile strength ▪ very high modulus ▪ good thermal and electrical conductivity ▪ brittle ▪ high dimensional stability, low thermal expansion

Figure 3.23 Structure and properties of carbon fibers (according to [3-17])

Figure 3.24 schematically shows the path from carbon fibers starting from the molecular structure via the microstructure, which resembles a rolled cigar, to the single filament and finally to the roving (also tow) as an endless multi-filament bundle, which is the usual delivery form for further processing.

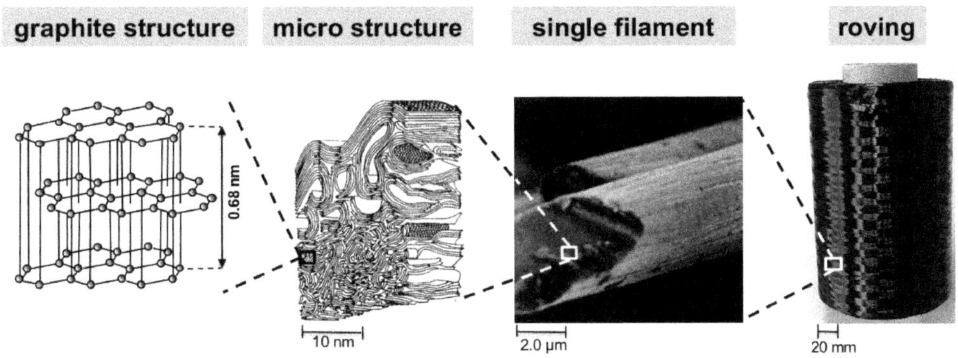

Figure 3.24 Structure of carbon fibers - scheme from microscopic to macroscopic (according to [3-17])

When using carbon fibers in unidirectional CFRP laminates, the tensile strength of carbon fibers is often the first consideration when comparing the properties of carbon fibers, although the compressive strength of the fiber is much more relevant for many applications. It is also usually assumed that the tensile strength of unidirectional CFRP laminates is dominated exclusively by the tensile strength of the fiber (see Figure 3.25). In fact, the tensile strengths of unidirectional CFRP laminates depend decisively on the elongation at break of the matrix system, at least in a certain stress range. As a rule of thumb, if the ratio of matrix elongation to fiber elongation is less than 2, the composite properties are matrix-dominated, while above this limit the fiber has a significant influence on the composite properties (see Figure 3.26) [3-19]. This also means that a higher matrix elongation (> 4%) would generally achieve better fiber utilization. This could be achieved, for example, by a targeted adaptation of the matrix or by the degree of curing of the matrix system. Since, however, the compressive strength of a composite often plays a significant role for the application, the increase of the matrix elongation is limited, since the compressive strength drastically decreases from a certain point of the fracture toughness increase (from approx. 4%); see Ref. [3-19]. The main cause of this phenomenon is, among other things, the greatly reduced supporting effect of the fibers due to the lower modulus of elasticity of ductile matrix systems. For thermoset systems, an elongation at break of around 3.5% or significantly higher is therefore targeted.

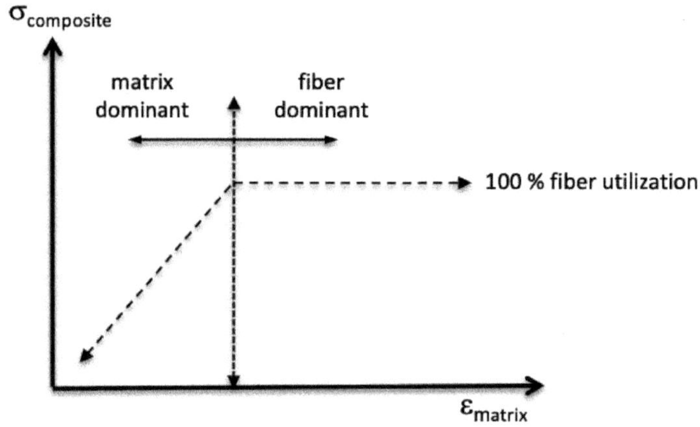

Figure 3.25 Schematic representation of the matrix and fiber-dominant tensile strength ranges of CFRP UD laminates (according to [3-19])

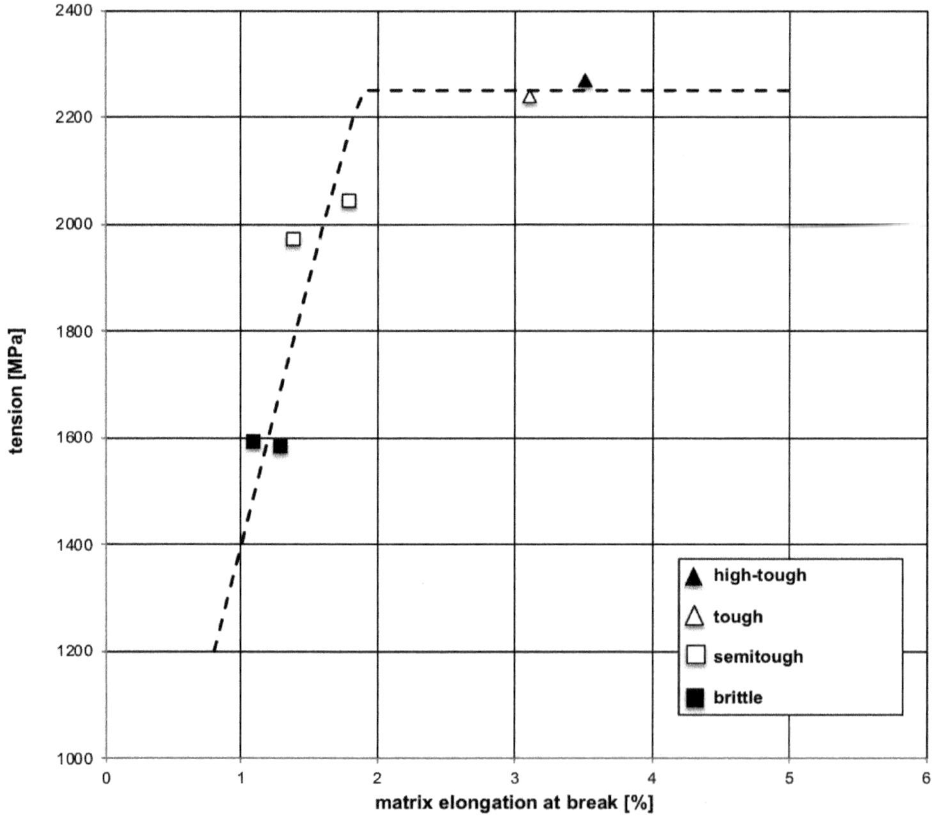

Figure 3.26 Tensile strength of unidirectional CF laminates vs. matrix fracture elongation (according to [3-19])

Table 3.4 gives an overview with typical properties of carbon fibers (PAN- or pitch-based). Besides the high level of mechanical properties, there are a number of other features of carbon fibers that make them attractive for use in a wide variety of applications. These include, for example, fatigue strength, damping behavior, corrosion resistance, and very low thermal expansion – all properties that are unique in this combination and predestine the material for highly stressed fiber composite components.

Table 3.4 General Properties of Carbon Fibers

Property Type	Characteristics
Mechanical properties	▪ high strength and stiffness (high modulus of elasticity) up to approx. 500 °C ▪ almost fatigue-free and fatigue-proof ▪ low creep tendency ▪ good vibration damping ▪ progressive stress-strain behavior
Chemical and physical properties	▪ chemically inert ▪ low density ▪ corrosion-resistant ▪ high resistance to acids, alkalis, and organic solvents
Thermal properties	▪ negative coefficient of expansion ▪ good to very good thermal conductivity
Electrical properties	▪ high electrical conductivity in fiber direction
Electromagnetic properties	▪ low absorption of X-rays ▪ non-magnetic

Table 3.5 shows examples of some of the above-mentioned properties compared with characteristic values of four different carbon fiber types based on PAN.

Table 3.5 Properties of Different Carbon Fiber Types [3-17]

Physical Property	Formula Symbol	Unit	HT (High-Strength)	IM (Intermediate)	HM (High Modulus)	HMS (High Modulus/ High Stiffness)
Density	ρ	g/cm³	1.74	1.80	1.83	1.85
Tensile strength	σ_B	MPa	3600	5600	2300	3600
Tensile modulus	E_Z	GPa	240	290	400	550
Compressive strength	σ_B	GPa	2.50	4.20	1.50	1.80
Elongation at break	ε_B	%	1.7–1.9	1.6–2.0	1.0–1.4	0.7–1.5
Fiber diameter	d	µm	7	5	6.5	5
Long-term temperature use	T_L	°C	500	500	500	500
Sublimation point	T_S	°C	3600	3600	3600	3600

It is easy to see that with an increase in the modulus of elasticity the strength and the filament diameter of the fibers decrease. In PAN-based fibers, fibers with a modulus of elasticity of < 300 GPa and a strength of up to about 4000 MPa are generally referred to as HT/HS fibers. IM fibers are typically characterized by a significantly higher strength of > 5000 MPa, with slightly higher stiffness (280 to 300 GPa) than HT/HS fibers. High modulus fibers (HM types) have a modulus of elasticity of > 300 GPa, with comparable tensile strength to HT fibers. However, the boundaries between fiber types are not precisely defined. Another very important characteristic value is elongation at break. Table 3.5 shows that the elongation at break can vary with the respective fiber type and that there are overlaps between the fiber types.

A very important characteristic of carbon fibers is the filament diameter or the so-called *yarn fineness*. The *yarn count* or the so-called *titer* (linear density) of a filament is defined as the length-related mass [3-16]. A high yarn count means a low ratio, and a high *titer* a higher ratio of mass (M) to length (L):

$$Titer = \frac{M}{L}[tex] \tag{3.1}$$

Internationally, the *titer* is measured in *tex*. This unit has replaced the old units *denier (den), metric number (Nm)* and *English number (Ne)*. In the textile industry, the old units are often still in use and can be converted as follows:

- 1 tex: 1 gram per 1000 meters
- 1 den: 1 gram per 9000 meters
- 1 Nm: 1 meter per 1 gram of yarn
- 1 Ne: 840 yards per English pound

Often the units carry multiplying prefixes such as deci or kilo:

- 1 dtex: 1 gram per 10,000 meters (decitex)
- 1 ktex: 1 kilogram per 1000 meters (kilotex)

Among each other, the units interconvert as follows:

- 1 tex = 9 den
- 1 tex = 1 g/1000 m = 1/1000 × 1 g/1 m = 1/1000 × 1/ Nm
- 1 tex = 590.5/Ne

With known density ρ of the fiber (for determination see Chapter 4), the fiber radius r can be determined from the titer using the following equation:

$$r = \sqrt{\frac{1}{\pi\rho} \cdot \frac{M}{L}} \; [\mu m] \tag{3.2}$$

Example: a carbon fiber filament has a mass M of 0.067 kg with a length L of 1000 m. The density ρ is approx. 1770 kg/m^3. This results in a radius r of about 3.5 µm. The theoretical diameter of this carbon fiber is therefore 7 µm (idealized). The calculation assumes a circular cross-section of the fiber in this case.

As described above in the manufacturing process, carbon fibers are produced in the form of thin, parallel filaments, which in turn are combined into thicker bundles known as rovings or tows. The cross-section of a roving is slightly elliptical or flat, depending on the application. There are also rovings with a slight protective twist (e.g. 10 twists/m), which makes the cross-section rounder and prevents the fibers from being broken up during weaving processes. Rovings are categorized according to the number of individual filaments and/or yarn count. The number of single filaments in the tow is described by the *K number* and given in full 1000s of filaments (1000 = K). Common delivery formats are:

- 1 K (1000 filaments) = 67 tex
- 3 K (3000 filaments) = 200 tex
- 6 K (6000 filaments) = 400 tex
- 12 K (12,000 filaments) = 800 tex
- 24 K (24,000 filaments) = 1600 tex

In this case, the *tex number* shown in the list refers to a filament diameter of 7 µm. If the titer changes due to a smaller or larger filament diameter, the tex number for the corresponding K number also changes. Thus, with a diameter of 5 µm and a K number of 1000, a titer of 44.5 tex would be obtained.

Rovings with a filament number of more than 24 K, e.g. 50 K, 100 K, or 400 K, are referred to as *heavy tows*, which must be spread accordingly for further processing, e.g. into textile fabrics.

The fineness of a tow depends on the number of individual filaments combined and their fineness and thus on the diameter and density of these filaments. A 12 K roving is given with a fineness of 800 tex and a 24 K roving with a fineness of 1600 tex (see above). In addition to material properties such as strength and processability, the titer has a significant influence on the price (see also Section 3.5.2).

The "inconspicuous" take-up winders for winding the carbon fiber yarn also comprise a very important part of the production process. These re-winders have a decisive influence on the yarn geometry, winding width, winding angle, and packing density of the bobbins. Whereas up to around 2010, for example, the yarn geometry still had a quasi-round cross-section ("laces") during winding, in the course of time the geometry became more and more elliptical, flat ribbons in order to be able to spread the yarn better in later process steps. Depending on the winder, the filament bundle is deposited in different widths. Another example is the winding angle, which is different for HT fibers and IM fibers. This can vary depending on the winder. The take-up winders have been improved considerably in recent years. Altogether, there are many versions and variants available on the world market.

The relationship between titer, fiber basis weight, and tow width has important consequences especially in the production of dry fiber tape (DFP tape, dry fiber placement). Common delivery forms of narrow DFP tapes are often ¼- and ½-inch, corresponding to 6.35 mm and 12.7 mm widths. Today these widths are based on prepreg slit-tapes, but will tend to become wider in future (e.g. 1.5 inches) in order to increase the laying performance of AFP machines. With a given width of the DFP tape and a given fiber type (and thus a fixed tex number), only certain fiber basis weights (FAW = fiber areal weight) can be realized. This is shown in the following examples of the DFP tape (HiTape®) from Hexcel for an AS7 or IMA fiber:

- Fiber type: HexTow® AS7, titer: 800 tex, tape width 6.35 mm, 1 roving, length 1000 m

$$FAW = \frac{800\ \text{tex}}{0.00635\ \text{m} \times 1000\ \text{m}} = \frac{800\ \text{g}\ (\text{in } 1000\ \text{m})}{6.35\ \text{m}^2\ (\text{in } 1000\ \text{m})} = 126\ \text{g}/\text{m}^2 \tag{3.3}$$

- Fiber type: HexTow® IMA, titer: 445 tex, tape width 6.35 mm, 2 rovings, length 1000 m

$$FAW = \frac{445\ \text{tex}}{0.00635\ \text{m} \times 1000\ \text{m}} = \frac{445\ \text{g}\ (\text{in } 1000\ \text{m})}{6.35\ \text{m}^2\ (\text{in } 1000\ \text{m})} \times 2\ \text{rovings} = 140\ \text{g}/\text{m}^2 \tag{3.4}$$

- For correspondingly higher fiber basis weights, two, three, or more tows must be used (Table 3.6). The same applies if the fiber basis weight to be processed is specified, which then determines the maximum width of the DFP tape (Table 3.7).

Table 3.6 DFP Tapes – Width vs. Fiber Weight per Unit Area (FAW)

Width in mm (inch)	HexTow® AS7 Fiber Basis Weight in gsm			HexTow® IMA Fiber Basis Weight in gsm				
6.35 (¼")	126	–	252	140	–	210	–	280
12.7 (½")	126	190	252	140	175	210	245	280

Table 3.7 DFP Tapes – Fiber Area Weight (FAW) vs. Width

Fiber basis weight in gsm	HexTow® AS7 Width in mm		HexTow® IMA Width in mm			
134	5.98	–	3.33	–	6.66	–
150	5.34	10.68	5.95	–	8.92	–
194	4.13	8.26	4.60	–	6.90	–
268	2.99	5.98	3.33	5.00	6.66	8.33

3.4.2 Carbon Fiber Types

With the continuous development of precursors, manufacturing processes, and their parameters, numerous new carbon fiber types have been developed over the decades since the first development and commercial production, but also many old types have disappeared from the market due to their properties and/or manufacturing costs. Today's fiber types, e.g. HT or IM fibers, offer about 10 to 15% more tensile strength and a 3 to 5% higher modulus of elasticity at about 10% lower precursor costs.

Table 3.8 shows examples of typical carbon fibers based on PAN, manufactured by companies such as *Tejin Carbon* and *Hexcel Composites,* as well as their properties.

Table 3.8 Properties of PAN-Based Carbon Fibers from *Tejin Carbon* and *Hexcel* [Sources: Data Sheets from *Tejin Carbon* and *Hexcel*]

class	fiber name	K value [-]	fiber diameter [µm]	tensile strength [MPa]	tensile modulus [GPa]	elongation at break [%]	tex [g/km]	density [g/cm³]	manufacturer
standard modulus	HTA40	1	7.0	4100	240	1.70	67	1.77	Toho Tenax
		3					200		
		6					400		
	HTS40	12	7.0	4400	240	1.80	800	1.77	
	AS4	3	7.1	4480	231	1.80	210	1.79	Hexcel
		6					427		
		12					858		
	AS7	12	6.9	4830	241	1.80	800	1.79	
	UTS50	12	6.9	5100	245	2.10	800	1.78	Toho Tenax
	ITS50	24	7.0	5100	265	1.90	1630	1.80	
intermediate modulus	IMS40	12	6.4	4500	290	1.60	680	1.73	Toho Tenax
	IMS60	12	5.0	5800	290	2.00	410	1.79	
	IM7	6	5.0	5175	276	1.87	223	1.78	Hexcel
		12		5520		2.00	446		
	IMA	12	5.1	6067	297	1.80	445	1.79	
	IM10	12	4.4	6964	310	2.00	324	1.79	
high modulus	HMA35	12	6.7	3300	355	0.90	760	1.79	Toho Tenax
	UMS40	12	4.8	4700	390	1.30	390	1.79	
	HM63	12	4.9	4688	441	1.00	420	1.83	Hexcel

In general, carbon fibers are classified according to their mechanical properties as follows:

Classification	Definition
HT/HS	High tenacity/strength
UHT/UTS	Ultra high tenacity/strength
AM	Advanced modulus
IM	Intermediate modulus
HM	High modulus
UHM	Ultra-high modulus
UMS	Ultra-modulus strength
HMS	High-modulus strength

As can be seen in Table 3.5 mechanical properties such as tensile strength and modulus of elasticity increase starting with the so-called standard fibers (HT/HS grades), through AM and IM fibers to high modulus fibers. Correspondingly, the filament diameter decreases and the elongation at break is reduced. HT and HS grades are well suited for the production of carbon fiber fabrics. Table 3.5 shows how the further development of HT standard fiber types (e.g. HTA and AS4 types) to HS fiber types (e.g. UTS50, AS7) has increased elongation at break and strength. These two properties are particularly interesting for unidirectional fiber applications such as UD prepregs. Figure 3.27 shows this connection. The advanced modulus type (AM type) C T50-4.8, which has only been available on the market since 2018, is remarkable for its even higher tensile strength at the level of IM types, combined with high rigidity and elongation at break.

For a simpler overview of the different fiber types Figure 3.28, shows a rough division of the fiber types into four broad fields. Thereby, the mechanical property levels of the respective fiber types are classified and compared to each other. The following classification will be used:

- Field 1 - HT fibers
- Field 2 - IM fiber
- Field 3 - HM fibers on PAN basis
- Field 4 - pitch-based HM fibers

The tensile strength decreases with increasing modulus of elasticity (stiffness). This applies in particular to the PAN-based *(field 3)* and the pitch-based HM types *(field 4)*. An exception is the IM fiber types *(field 2)*, which have a considerably higher tensile strength than HT standard fibers *(field 1)* despite an increased modulus of elasticity.

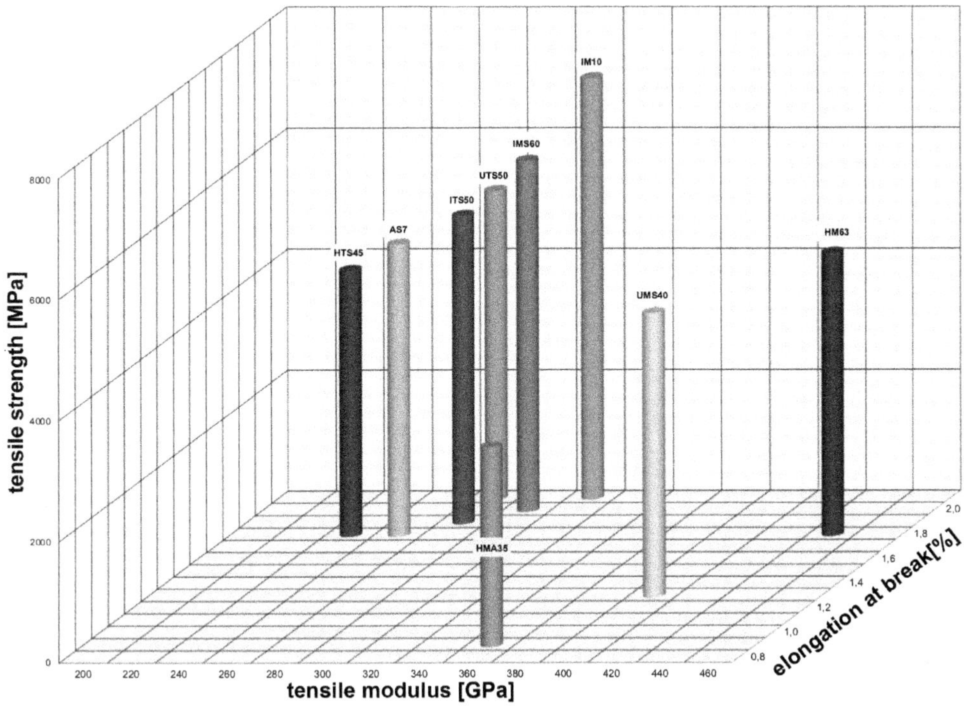

Figure 3.27 Comparison of Mechanical Properties of PAN-Based Carbon Fibers

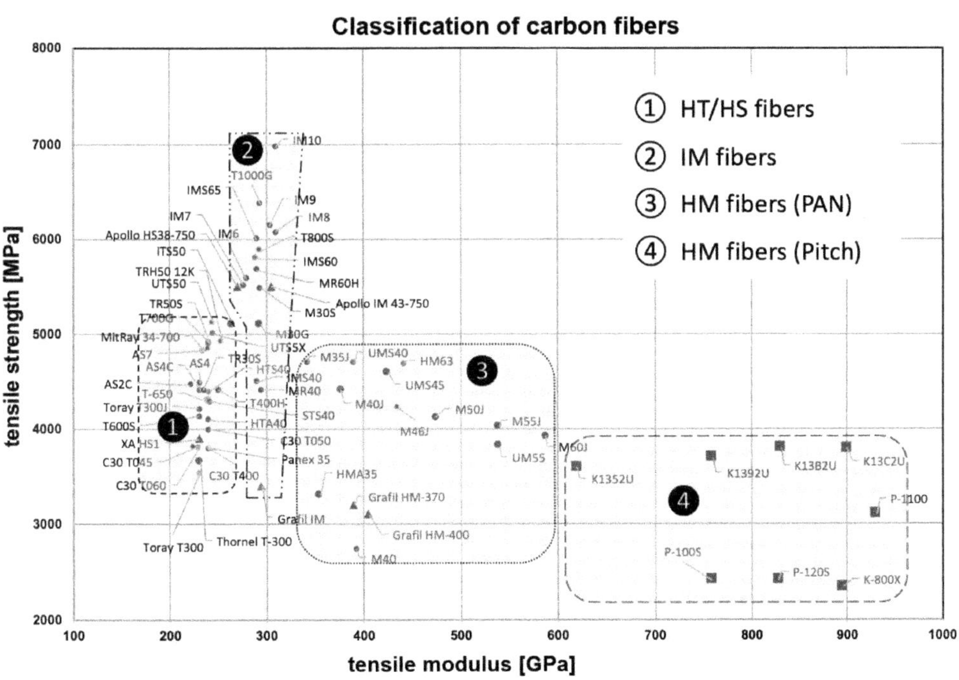

Figure 3.28 Classification of carbon fiber types

The graph in Figure 3.28 also shows that there is a strong concentration of many different PAN-based fiber types in the HT/HS and IM fiber range *(fields 1 & 2)*. As a result, commercially large quantities of these fibers are available and, at the same time, the price level in the respective segment is relatively low. The trend towards reducing costs at this property level continues (see also Chapter 7). Accordingly, the prices are considerably higher (up to 1000 €/kg (ca. 1150 US$/kg)) for less in-demand fibers with special up to very extreme properties (for pitch-based types).

The graphic in Figure 3.29 gives a detailed comparison overview of the mechanical properties of selected commercially available and historical fiber types. The square markings (■) show pitch-based carbon fibers and the round markings (●) show PAN-based carbon fibers. As an example of historical fiber types from the 1980s, those of *Courtaulds* are shown in the diagram and marked with a triangle (▲). *Courtaulds* was together with *Toho* (today: *Tejin Carbon*) the first company in Europe to produce carbon fibers. Due to the large number of fibers available today, not all types and manufacturers have been included in the graphic for a better overview. In addition to Figure 3.29, Table 3.13 at the end of the chapter contains a detailed tabular overview with manufacturers, fiber types, and some selected properties.

A closer look at Figure 3.29 reveals that today there are HT fiber types (e. g. ITS50 and UTS50) which have already exceeded the limit of IM fibers according to the criteria mentioned in Section 3.4.1 regarding their tensile strength (> 5000 MPa), and have almost reached it regarding the Young's modulus (280 to 300 GPa).

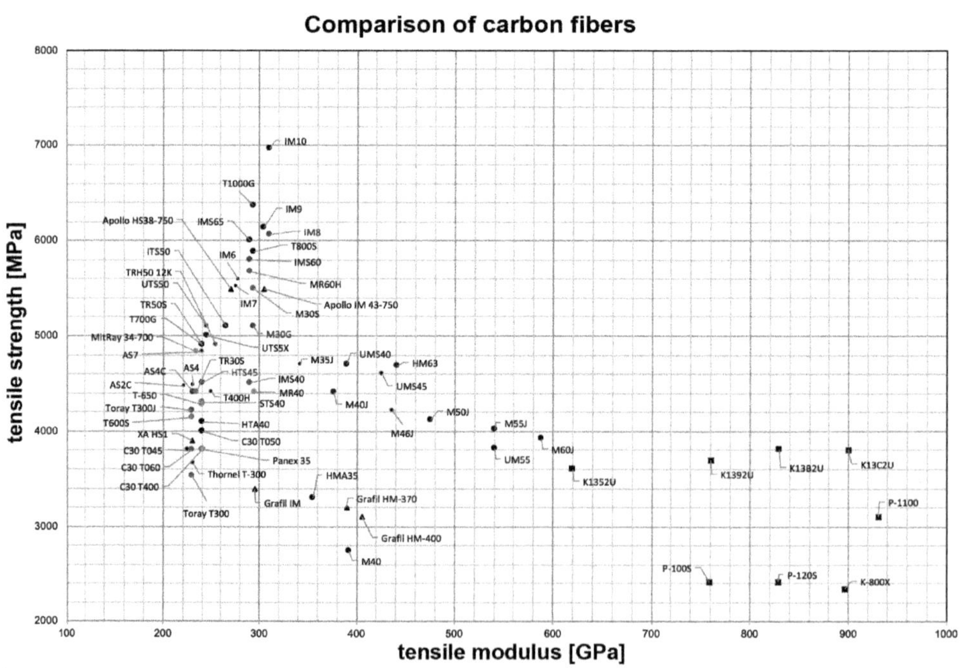

Figure 3.29 Comparison of commercial carbon fibers from different manufacturers

Figure 3.30 shows a comparison of PAN-based carbon fibers, aramid and glass fibers, as well as metallic materials. The higher modulus of elasticity and the significantly higher tensile strength of all carbon fiber types compared to the other materials are clearly visible. Only glass and aramid fibers have a tensile strength close to that of HT carbon fibers. At the same time, it should not be overlooked that carbon and aramid fibers, unlike metals and the fiber composite components made from them, have anisotropic properties, which require special consideration in the design of components and the processing of these semi-finished fibers.

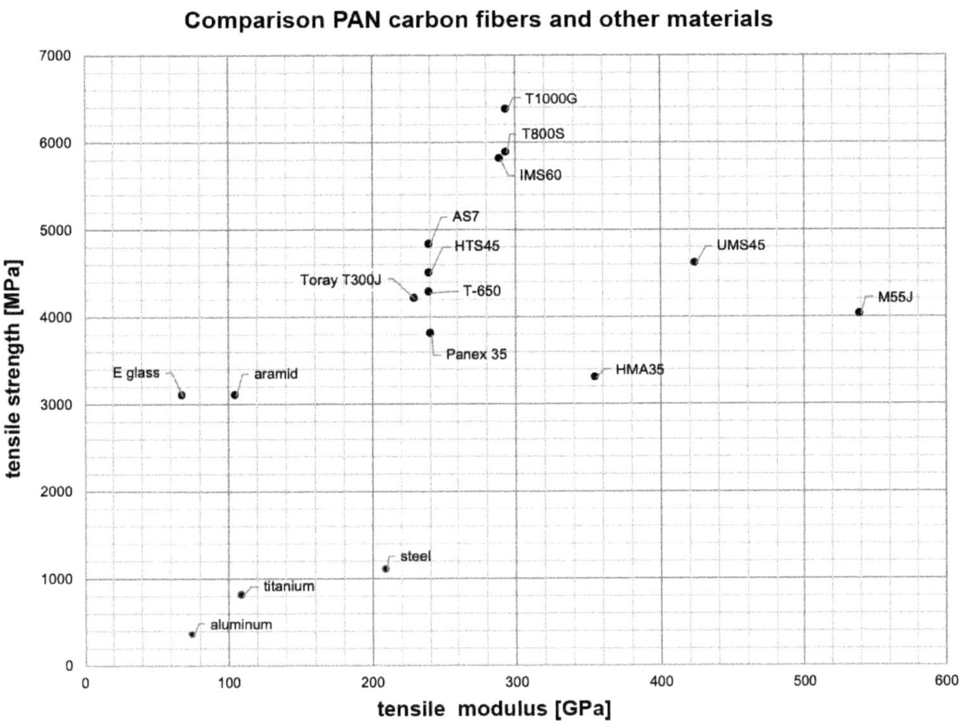

Figure 3.30 Comparison of properties of PAN carbon fibers with other materials

3.5 Manufacturers and Costs

This section starts with a brief overview of the precursor and carbon fiber manufacturers available on the world market and continues with explaining the manufacturing costs and their background. At the end of the chapter, an overview of the currently available production capacities for carbon fibers is given.

3.5.1 Manufacturers of Precursors and Carbon Fibers

The total number of carbon fiber manufacturers has remained relatively constant over the last 20 years, despite increasing demand and use of carbon fibers, with about 20 manufacturers active (see also Ref. [3-25]). During this time, smaller manufacturers in larger groups in Europe and the USA - often retaining the fiber type designation - have been bought up, and new manufacturers have joined the ranks, especially in Asia, India, and Russia. At the same time, the increased demand is reflected in the increased production capacities available on the world market (see Section 3.5.3, Figure 3.34).

Table 3.9 provides an overview of current global suppliers of PAN- and pitch-based carbon fibers (as of June 2020).

Table 3.9 Global Overview Carbon Fiber Manufacturers

Fiber Type	Manufacturer	PAN Production	Fiber Manufacturing	Brand Name	Full Assortment
PAN based	*Toray*	Japan, USA, France	Japan, USA, France	TORAYCA	x
	Tejin Carbon	Japan	Germany, Japan, USA	TENAX	x
	Mitsubishi	Japan	USA, Japan	PYROFIL, GRAFIL	x
	FORMOSA	Taiwan	Taiwan	TAIRYFIL	
	Solvay Cytec	USA	USA	THORNEL	
	Hexcel	USA, France	USA, Spain	HEXTOW	x
	ZOLTEK	Hungary, Mexico	USA, Hungary, Mexico	PANEX	
	SGL	Japan	USA, Ireland	SIGRAFIL	
	DowAksa	Turkey	Turkey	AKSA	
	UMATEX	Russia	Russia	UMT	
	Kemrock	India	India	JAITEC	
	JIANGSU HENGSHEN FIBRE	China	China	n.a.	x
	Dalian Xingke Carbon Fiber Co	China	China	n.a.	
	GanSu HaoShi Carbon Fiber Co	China	China	n.a.	
	Zhongfu Shenying	China	China	SYT45	

Fiber Type	Manufacturer	PAN Production	Fiber Manufacturing	Brand Name	Full Assortment
Pitch based	*Hyosung Corporation*	South Korea	South Korea	TANSOME	
	LeMond Composites	USA	USA	GRAIL	
	Mitsubishi Chemical	Japan	USA, Japan	DIALED	
	Solvay Cytec	USA	USA	THORNEL	

The full-range suppliers (see Table 3.9, right column) are those who offer a wide range of different titers (3, 6, 12, 24, 48 K) and PAN-based fiber types (HT, IM, HM ...) or are technically capable of doing so. In the last 10 years, several new suppliers of carbon fibers of different qualities and titers have been added, especially on the Asian market. In particular, *Jiangsu Hengshen* can compete with established manufacturers due to its range of products.

Table 3.10 provides an overview of current precursor manufacturers on the world market, some of whom resell the PAN precursor to carbon fiber manufacturers that do not develop and manufacture their own PAN. In addition to a few producers, such as *Mitsubishi*, some new, partly pure PAN producers (e.g. *Bluestar*, *Solvay*) have been added, which have penetrated into the carbon fiber market, for example by further developing textile polyacrylonitrile, which could only be used for the clothing industry. Both process steps - production of a suitable PAN precursor and conversion to a competitive carbon fiber - require a high degree of technical and chemical know-how as well as considerable investment, so that today the complete process chain is not always mapped by one manufacturer.

Table 3.10 Overview of Precursor Manufacturers and Suppliers on the World Market

- *Bluestar Fibers (Lanzhou)*
- *SGL Carbon*
- *Solvay (until 2017: European Carbon Fiber GmbH)*
- *Chemtex International Inc.*
- *DowAksa*
- *Mitsubishi Rayon Co. Ltd.*
- *FISIPE SA*
- *FRP Services & Co.*
- *Kemrock Industries & Exports Ltd.*
- *Composite Technical Services LLC*
- *Omnia LLC*
- *Tango Engineering Ltd.*

Long-established carbon fiber manufacturers on the market, such as *Tejin Carbon, Hexcel,* and *Toray,* have consistently advanced the expansion of their own PAN capacities in recent years in order to achieve the strongest possible vertical integration of the entire production chain from precursor to finished semi-finished product (see also Section 3.5.3).

3.5.2 Manufacturing Costs

The most widely used precursors for the production of carbon fibers today are polyacrylonitrile (PAN) based. Looking at the manufacturing costs for carbon fibers (Figure 3.31), it becomes very clear that more than 50% of the costs are attributable to the precursor. The remaining 49% are distributed among oxidative stabilization (15%), carbonization/graphitization (23%), and preparation application (sizing) and wrapping (11%). The majority of the costs after precursor production are attributable to stabilization and carbonization, since the mass loss during these two process steps is around 50% for PAN and 60 to 75% for pitch, which makes the latter fibers even more expensive than PAN. Potential cost reduction are essentially connected to the production of precursors or the starting material for the precursor.

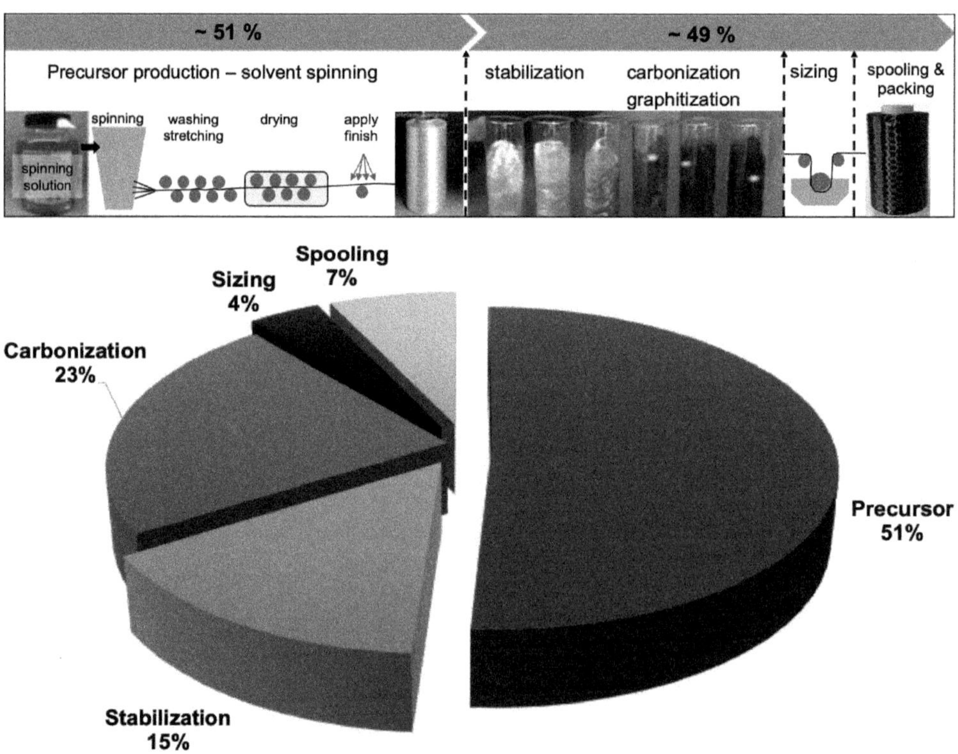

Figure 3.31 Cost allocation for carbon fiber production based on PAN [3-24]

In addition, precursor production based on pitch already requires a higher energy demand than PAN fibers, which results in a high carbon footprint. In addition to their specific properties, both aspects have made the broad use of pitch-based fibers less attractive, so that these fiber types are only used for special applications.

The production costs for carbon fibers have been reduced steadily over the past decades which results in increasing application and the construction of extensive production capacities. However, since the two most important precursors (PAN, pitch) are obtained from crude oil derivatives, a price increase for these and all other alternative crude oil-based precursors can be assumed in the event of rising crude oil prices.

As already mentioned above in Section 3.4.1 the titer has a significant influence on the price in addition to material aspects such as strength and processability. The finer the yarn or the lower the number of filaments in the bundle *(K number)*, the higher the price, since less fiber can be produced at the same throughput speed. In addition, due to the reduced diameter, the precursor yarn for IM and HM types reacts much more sensitively to the temperature employed in the oxidation stage. Therefore, the throughput speed must be reduced by up to 50% in order to prevent thermal damage to the precursor. In summary, the price level is essentially determined by the fiber diameter (titer) and the fiber properties (see Figure 3.33).

Due to the continuous further development and improvement of the manufacturing processes since the mid-1980s, fibers with up to 480 K can be produced today. The further developments will focus on the precursors, the properties of which could be increasingly adapted to the processes. Process improvements include, for example, in the conversion from vertical furnaces, in which the fibers were guided vertically through the oxidation stage and carbonization, to horizontal furnaces. The latter offer advantages in terms of heat distribution and control as well as the possibility of extension (lengthening of the line).

The maximum throughput of a fiber line is determined by several factors. The width of the fiber lines limits the amount of carbon fiber that can be produced per time unit. This fixed width also determines the maximum amount of fiber material (K number) per unit time that can be converted in the PAN stabilization stage (oxidation stage), so that unfortunately the costs are not inversely proportional to the K number. Particularly in pre-carbonization, the balance between the thermal energy supplied to the precursor and the amount of heat released by the precursor during the conversion processes must be maintained. Since considerable amounts of conversion energy may be released depending on the precursor, it must be possible to control and dissipate these in a targeted manner to prevent burning. Depending on the plant design, however, there are limits to this, so the titer cannot be increased arbitrarily.

With today's modern fiber lines, the processes are automated and optimized to such an extent that the costs for operating a carbonation plant can hardly be reduced significantly. These are essentially determined by the energy costs for plant operation and development costs for precursors, and in the case of aeronautically qualified fiber types additionally by the certification costs. This also means that the cost per kg of fiber is almost independent of the sales volume once a fiber line is in continuous operation. In fact, a 12 K HT fiber qualified for aviation cost about 56 US$/kg 30 years ago. Today the price is around 34 US$/kg, i.e. the improvements and developments mentioned above have contributed to the fact that the fiber price has remained very stable over time despite increased operating and energy costs. The sharp increase in the use of carbon fibers has also contributed to this, as a result of which numerous new lines have been set up, so that the possibility of "single-variety" operation has been created and unproductive set-up times have been significantly reduced. It can be assumed that the current fiber production facilities have a capacity utilization of > 90%.

Figure 3.32 and Figure 3.33 visualize the relationship between fiber properties and fiber price. Figure 3.32 shows how the price level changes with the strength of the carbon fiber. For comparison, metallic materials such as steel, aluminum, and titanium as well as aramid and E-glass fibers are also shown. The diagram clearly shows that the price level within a fiber class (HT fiber aerospace vs. HT fiber industry) changes only slightly with approximately the same strength. This is related to the capacity and operating costs of a fiber line. The diagram in Figure 3.33 shows – as already mentioned above in the text – a significant dependence of the fiber price on the titer. The lower the titer (and filament diameter), the higher the fiber price per kilogram of fiber.

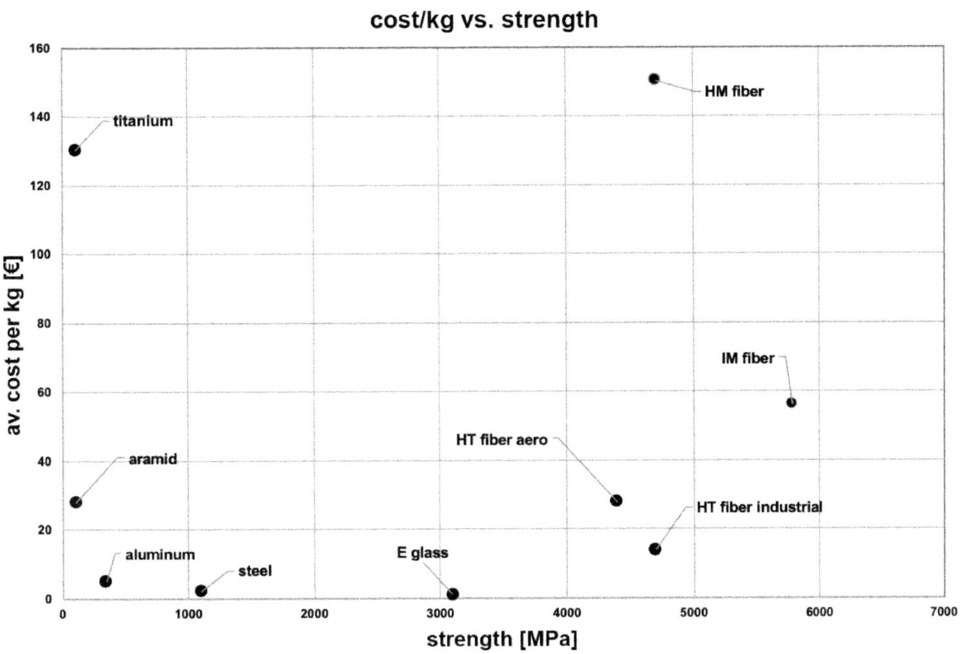

Figure 3.32 Relationship between fiber price and fiber strength

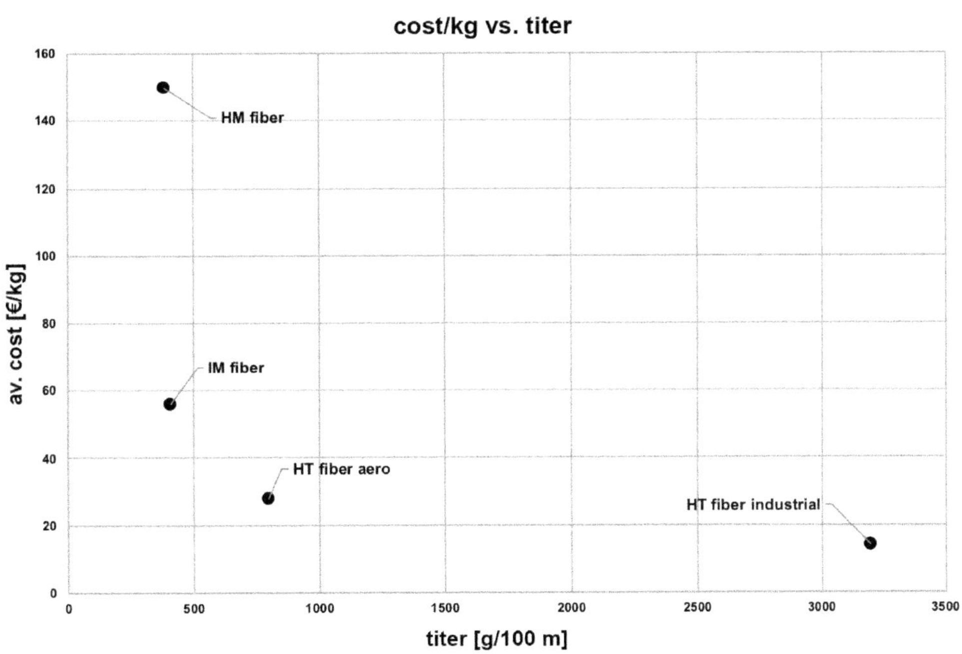

Figure 3.33 Relationship between fiber price and titer

3.5.3 Manufacturing Capacities

Figure 3.34 provides an overview of the theoretical global production capacities for carbon fibers. These add up to around 151,000 tons, whereby the figures may vary depending on the source. Compared to the theoretical quantities of approximately 14,000 t [3-19] available in the year 2000, this means an over tenfold increase in capacities [3-20] [3-24].

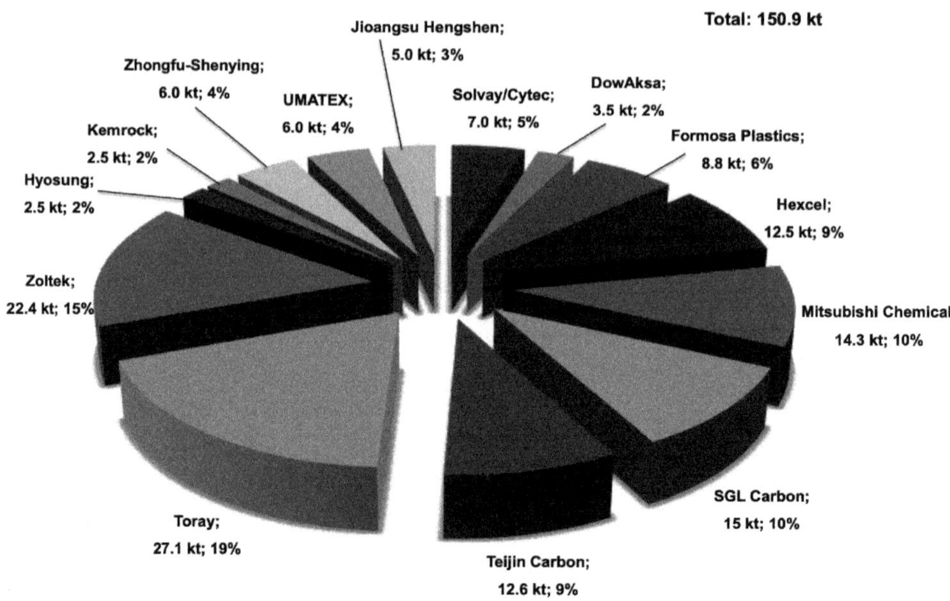

Figure 3.34 Annual theoretical production capacity for carbon fiber, as of 09/2019 (according to [3-20] [3-24])

These figures, based on manufacturer data, express only the theoretically producible quantity, not the actual world market demand for carbon fibers. No distinction is made between different fiber types, K number, or application areas (aerospace or industrial fibers). In Table 3.11 the absolute and percentage capacities per manufacturer are listed. Figure 3.35 illustrates the global distribution of capacity in a country comparison.

The chart in Figure 3.35 shows the theoretical production capacity broken down by country of manufacture. Three major regions account for more than 60% of capacity: North and Central America (38%), Japan (18%), and China (11%).

Table 3.11 Manufacturer, Brand Name, and Theoretical Manufacturing Capacities, as at 09/2019

Manufacturer	Type Designation	Capacity in kt	Capacity in %
Solvay (Cytec)	Thornel®	7.0	4.6
DowAksa	not specified	3.5	2.3
Formosa Plastics	Tairyfil	8.8	5.8
Hexcel	Magnamite, HexTow®	12.5	8.3
Mitsubishi Chemical	DIALED®, Pyrofil, Grafil	14.3	9.5
SGL Carbon	SIGRAFIL®	15.0	9.9
Tejin Carbon	Tenax®-E, Tenax®-J	12.6	8.3
TORAY	TORAYCA®	27.1	18.0
Zoltek	PANEX®	22.4	14.8
Hyosung	TANSOME®	2.5	1.7
Kemrock	JAITEC®	2.5	1.7
Zhongfu Shenying	SYT45	6.0	4.0
UMATEX	UMT®	6.0	4.0
Jiangsu Hengshen	not specified	5.0	3.3
LeMond Composites	GRAIL™	–	0.0
Other	not specified	5.7	3.8
	Total	**150.9**	**100**

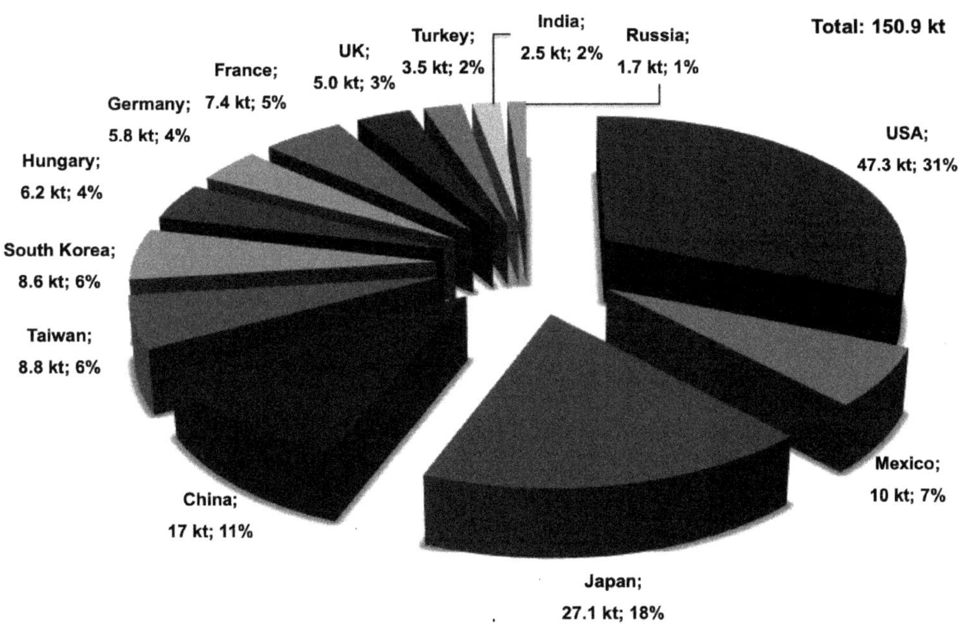

Figure 3.35 Theoretical production capacity by country of manufacture, as at 09/2019 (according to [3-20])

The theoretical production capacities are offset by the global demand for carbon fibers. Here, volumes have more than doubled since 2010. An increase to more than 100,000 t/a by 2021 is expected, which would roughly triple the demand since 2010 [3-20]. In addition to the annual development of total global demand, Figure 3.36 shows the forecast of a percentage increase up to 2023. An annual increase of around 11% is expected [3-20].

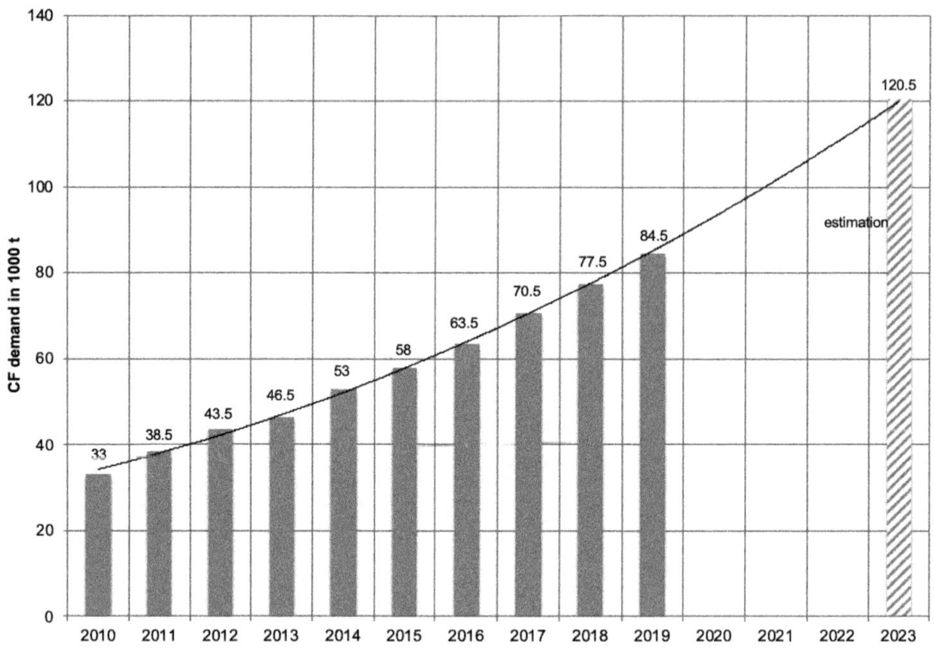

Figure 3.36 Global carbon fiber demand, status 09/2019 (after [3-20])

While production capacities are currently still predominantly outside Europe, the chart in Figure 3.37 shows that Europe has the largest annual demand of 26% after North America (34%), followed by Asia (24%) and Japan (11%) (as of 09/2019, [3-20] [3-24]). New plants for PAN and carbon fiber production will be built in Europe due to the forecast increase in demand. Some important manufacturers like *Tejin Carbon* (Germany) as well as *Toray* and *Hexcel* (France) have already started or completed their first expansions.

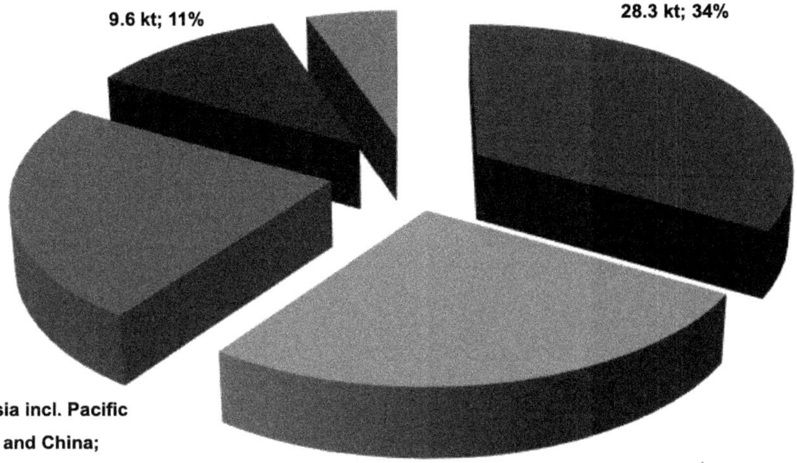

Figure 3.37 Global, annual demand for carbon fibers by region, as of 09/2019 [3-20]

3.5.4 Suppliers of Process Technologies

Table 3.12 provides an exemplary overview of manufacturers on the world market who offer precursors and process technology for the production or further processing of precursors and carbon fibers.

Table 3.12 Providers of Process Technologies [3-24]

Technology	Supplier
▪ Oxidation and carbonation furnaces ▪ Carbon fiber lines ▪ Precursor and carbon fiber lines	▪ *Eisenmann* Anlagenbau *GmbH* ▪ *Harper International Inc.* ▪ *Trützschler Ltd.* ▪ *EPC Group* ▪ *Izumi International Inc.*
▪ Take-up winders for fiber lines	▪ *Oerlikon Barmag* ▪ *Georg Sahm GmbH & Co. KG* ▪ *Kamitsu Seisakusho Ltd.* ▪ *Izumi International Inc.* ▪ *AIKI RIOTECH Corporation*

Table 3.12 Providers of Process Technologies [3-24] *(continued)*

Technology	Supplier
▪ Bobbin creels for fiber and prepreg lines	▪ *Izumi International Inc.* ▪ *Texmer GmbH* ▪ *Cygnet Texkimp Ltd.* ▪ *KARL MAYER* Textilmaschinenfabrik *GmbH* ▪ *ATA Carbon Fiber Tech.* ▪ *Van Wees* ▪ *UD and Crossply Technology B. V.* ▪ *CT Platon*

Outlook

In the following chapters, the testing of the material properties of carbon fibers in the sense of property determination or quality assurance during the manufacturing process (Chapter 4) will be shown first. Chapter 5 provides numerous examples concerning the use of carbon fibers in the various forms (fiber flour, short fibers, and continuous yarns), Chapter 6 presents recycling processes and reuse of carbon fiber. Chapter 7 concludes with a brief overview of new developments in precursors and carbonization processes.

References for Chapter 3

[3-1] I. L. Shabalin, *Ultra-High Temperature Materials - Carbon (Graphene/Graphite) and Refractory Metals*, 2014, ISBN 978-94-007-7586-2

[3-2] T. Edison, US 470925, 1880-03-15

[3-3] J. Chen, C. Wang, X. Dong, H. Liu, *Study on the Coagulation Mechanism of Wet-Spinning PAN Fibers*. Journal of Polymer Research, 2006, 13, pp. 151-519

[3-4] R. Moreton, W. Watt, *The spinning of polyacrylonitrile fibres in clean room conditions for the production of carbon fibres*. Carbon, 1974, 12, pp. 543-554

[3-5] D. R. Paul; *A study of spin ability in the wet-spinning of acrylic fibers*. Appl. Polym. Sci., 1968, 12, pp. 2273-2298

[3-6] D. R. Paul; *Diffusion during the coagulation step of wet-spinning*. Appl. Polym. Sci., 1968, 12, pp. 383-402

[3-7] G. East, J. Mcintyre, G. Patel: *The dry-jet wet spinning of an acrylic fiber yarn*. J. Text. Inst., 1984, 75, pp. 196-200

[3-8] H. C. Hyang, K. Bo-Hye, H. L. Sung, Y. S. Kap, *Preparation of Carbon Fiber from Melt Spinnable PAN Co-polymer*. Journal of the Korean Chemical Society, 2013, 57, pp- 289-294

[3-9] K. Morita, Y. Murata, A. Ishitani, K. Murayama, T. Ona, A. Nakajima, *Characterisation of commercially available PAN (Polyacrylonitrile)-based carbon fibers*. Pure Appl. Chem., 1986, 58, pp. 455-468

[3-10] G. H. Olive, S. Olive, *Intermolecular nitrile reaction between paired dipoles*. Polymer Bulletin, 1981, 5, pp. 457-469

[3-11] T. Cho, Y. S. Lee, R. Rao, A. M. Rao, D. D. Edie, A. A. Ogale, *Structure of carbon fiber obtained from nanotube-reinforced mesophase pitch*. Carbon, 2003, 41, pp. 1419-1429

[3-12] R. C. Houtz, *"Orlon" Acrylic Fiber: Chemistry and Properties.* Textile Research Journal, November 1950, 20, pp. 786-801

[3-13] E. Fitzer, D.J. Müller, *The influence of oxygen on the chemical reactions during stabilization of PAN as carbon fiber precursor.* Carbon, 1975, 1, pp. 63-69

[3-14] E. Fitzer, W. Frohs, M. Heine, *Optimization of stabilization and carbonization treatment of PAN fibers and structural characterization of the resulting carbon fibres.* Carbon, 1986, 4, pp. 387-395

[3-15] H. Jäger, T. Hauke, *Carbonfasern und ihre Verbundwerkstoffe: Herstellungsprozesse, Anwendungen und Marktentwicklung.* Süddeutscher Verlag onpact; Stuttgart 2010

[3-16] Lumitos AG: *http://www.chemie.de/lexikon/Garnfeinheit.html*, accessed Dec. 19, 2017

[3-17] *Kohlenstofffasern*, taken from "Handbuch Faserverbundkunststoffe - Grundlagen, Verarbeitung, Anwendungen", AVK - Industrievereinigung Verstärkte Kunststoffe e.V., Bernd Wohlmann, 2009, 3rd edition

[3-18] C. Cherif (ed.), *Textile Materials for Lightweight Construction - Techniques - Processes - Materials - Properties*, Springer-Verlag, Berlin/Heidelberg, 2016, ISBN 978-3-662-46340-6, p. 114

[3-19] M. Flemming, G. Ziegmann, S. Roth, *Faserverbundbauweisen - Fasern und Matrix*, Springer-Verlag, Berlin/Heidelberg 1995, ISBN 3-540-58645-8

[3-20] M. Kühnel, M. Sauer, Carbon Composites e.V.: *The global CF and CC market 2019*, International Composites Congress (ICC) Stuttgart, September 2019

[3-21] *Ammonoxidation*: Industrial Organic Chemistry, Klaus Weissermel, Hans-Jürgen Arpe, John Wiley & Sons, 3rd edition, 1997, ISBN 3-527-28838-4

[3-22] H. Mainka, O. Träger, O. Stoll, E. Körner, A.S. Herrmann, *Alternative Precursors for Sustainable and Cost-Effective Carbon Fibers usable within the Automotive Industry.* Society of Plastic Engineers (Automotive Division) - Automotive Composites Conference & Exhibition 2013, Novi, Mich. USA

[3-23] M. Flemming, S. Roth, *Faserverbundbauweisen - Eigenschaften*, Springer-Verlag, Berlin/Heidelberg 2003, ISBN 3-540-00636-2

[3-24] Authors' market research on manufacturers of CF capacities on the world market

[3-25] G. Xian, Z. Wang, *Carbon Fiber Reinforced Plastics - Properties*, Comprehensive Composite Materials II, Volume 2, Elsevier Ltd. 2018, DOI: 10.1016/B978-0-12-803581-8.09246-8

[3-26] F. Tanaka, T. Okabe, *Historical Review of Processing*, Microstructures, and Mechanical Properties of PAN-Based Carbon Fibers, Comprehensive Composite Materials II, Volume 1, Elsevier Ltd. 2018, DOI: 10.1016/B978-0-12-803581-8.09895-7

Other Sources

Material data sheets of the fiber manufacturers

Composites World online magazine *(www.compositesworld.com)*

JEC Composites online magazine *(www.jeccomposites.com)*

Günter Niederstadt: *Ökonomischer und ökologischer Leichtbau mit faserverstärkten Polymeren*, Expert-Verlag, 2nd edition, 1997, ISBN 3-8169-1416-0

Table 3.13 Supplementary Table to Figure 3.29: Manufacturer, Brand Name, Fiber Designation, and Properties

Manufacturer	Brand Name	Fiber Name	Number of Filaments in K	Linear Density in tex	Diameter in µm	Tensile Strength in MPa	Tensile Modulus in GPa	Elongation in %	Density in g/cm³	Fiber Type/ Remark
Hexcel	HexTow	AS2C	3	200	6.9	4470	221	1.9	1.80	HT
Hexcel	HexTow	AS2C	6	400	6.9	4470	221	1.9	1.80	
Hexcel	HexTow	AS2C	12	800	6.9	4470	221	1.9	1.80	
Hexcel	HexTow	AS4	3	210	7.1	4619	231	1.8	1.79	
Hexcel	HexTow	AS4	6	427	7.1	4413	231	1.7	1.79	
Hexcel	HexTow	AS4	12	858	7.1	4413	231	1.7	1.79	
Hexcel	HexTow	AS4A	12	858	7.1	4412	231	1.8	1.79	
Hexcel	HexTow	AS4C	3	200	6.9	4654	231	1.8	1.78	
Hexcel	HexTow	AS4C	6	400	6.9	4447	231	1.7	1.78	
Hexcel	HexTow	AS4C	12	800	6.9	4482	231	1.8	1.78	
Hexcel	HexTow	AS4D	12	765	6.7	4862	241	1.8	1.79	
Hexcel	HexTow	AS7	12	800	6.9	4895	248	1.7	1.79	
Hexcel	HexTow	IM2A	12	446	5.2	5309	276	1.7	1.78	IM
Hexcel	HexTow	IM2C	12	446	5.2	5723	296	1.8	1.78	
Hexcel	HexTow	IM4	12	723	6.6	4796	276	1.7	1.78	
Hexcel	HexTow	IM5	12	700	6.5	5343	276	1.8	1.79	
Hexcel	HexTow	IM6	12	446	5.2	5723	279	1.9	1.76	
Hexcel	HexTow	IM7	6	223	5.2	5516	276	1.9	1.78	
Hexcel	HexTow	IM7	12	446	5.2	5654	276	1.9	1.78	
Hexcel	HexTow	IM8	12	446	5.2	6067	310	1.8	1.78	
Hexcel	HexTow	IMA	12	445	5.1	6067	297	1.8	1.79	
Hexcel	HexTow	IM9	12	335	4.4	6136	304	1.9	1.80	
Hexcel	HexTow	IM10	12	324	4.4	6964	310	2.0	1.79	
Hexcel	HexTow	HM63	12	420	4.9	4688	441	1.0	1.83	CAR-BIDE
Tejin Carbon	Tenax-J	HTA40	1	67	7.0	4100	240	1.7	1.77	HT
Tejin Carbon	Tenax-E	HTA40	3	200	7.0	4100	240	1.7	1.77	
Tejin Carbon	Tenax-E	HTA40	6	400	7.0	4100	240	1.7	1.77	
Tejin Carbon	Tenax-J	HTS40	3	200	7.0	4400	240	1.8	1.77	

Manufacturer	Brand Name	Fiber Name	Number of Filaments in K	Linear Density in tex	Diameter in µm	Tensile Strength in MPa	Tensile Modulus in GPa	Elongation in %	Density in g/cm³	Fiber Type/ Remark
Tejin Carbon	Tenax-J	HTS40	6	400	7.0	4400	240	1.8	1.77	HT
Tejin Carbon	Tenax-E	HTS40	12	800	7.0	4400	240	1.8	1.77	
Tejin Carbon	Tenax-E	HTS40	24	1600	7.0	4400	240	1.8	1.77	
Tejin Carbon	Tenax-E	HTS45	12	800	7.0	4500	240	1.9	1.77	
Tejin Carbon	Tenax-J	HTR40	24	1550	7.0	4482	235	1.9	1.84	
Tejin Carbon	Tenax-J	STS40	24	1600	7.0	4300	240	1.8	1.78	
Tejin Carbon	Tenax-J	STS40	48	3200	7.0	4300	250	1.7	1.77	
Tejin Carbon	Tenax-J	UTS50	12	800	7.0	5100	245	2.1	1.78	
Tejin Carbon	Tenax-J/E	UTS50	24	1600	7.0	5100	245	2.1	1.78	
Tejin Carbon	Tenax-E	ITS50	24	1600	7.0	5100	265	1.9	1.80	
Tejin Carbon	Tenax-J	IMS40	3	170	7.0	4710	294	1.6	1.74	IM
Tejin Carbon	Tenax-J	IMS40	6	340	7.0	4710	294	1.6	1.74	
Tejin Carbon	Tenax-J	IMS40	12	670	7.0	4710	294	1.6	1.74	
Tejin Carbon	Tenax-J	IMS60	24	830	5.0	5800	290	2.0	1.79	
Tejin Carbon	Tenax-J	IMS65	24	830	5.0	6000	290	2.1	1.78	

Table 3.13 Supplementary Table to Figure 3.29: Manufacturer, Brand Name, Fiber Designation, and Properties *(continued)*

Manufacturer	Brand Name	Fiber Name	Number of Filaments in K	Linear Density in tex	Diameter in µm	Tensile Strength in MPa	Tensile Modulus in GPa	Elongation in %	Density in g/cm³	Fiber Type/ Remark
Tejin Carbon	Tenax-J	UMS40	24	800	4.9	4700	390	1.2	1.79	CAR-BIDE
Tejin Carbon	Tenax-J	UMS45	12	385	4.7	4600	425	1.1	1.83	
Tejin Carbon	Tenax-J	UT800	24	905	5.5	5390	255	2.1	1.80	
Tejin Carbon	Tenax-A	HMA35	12	770	6.7	3240	345	0.9	1.79	
Tejin Carbon	Tenax-J	UM55	12	360	4.4	3820	540	0.7	1.92	
Toray	Torayca	T300	1	66	n.a.	3530	230	1.5	1.76	HT
Toray	Torayca	T300	3	198	n.a.	3530	230	1.5	1.76	
Toray	Torayca	T300	6	396	n.a.	3530	230	1.5	1.76	
Toray	Torayca	T300	12	800	n.a.	3530	230	1.5	1.76	
Toray	Torayca	T300B	1	66	n.a.	3530	230	1.5	1.76	
Toray	Torayca	T300B	3	198	n.a.	3530	230	1.5	1.76	
Toray	Torayca	T300B	6	396	n.a.	3530	230	1.5	1.76	
Toray	Torayca	T300B	12	800	n.a.	3530	230	1.5	1.76	
Toray	Torayca	T400HB	3	198	n.a.	4410	250	1.8	1.80	
Toray	Torayca	T400HB	6	396	n.a.	4410	250	1.8	1.80	
Toray	Torayca	T700SC	12	800	n.a.	4900	230	2.1	1.80	
Toray	Torayca	T700SC	24	1650	n.a.	4900	230	2.1	1.80	
Toray	Torayca	T800SC	24	1030	n.a.	5880	294	2.0	1.80	IM
Toray	Torayca	T800HB	6	223	n.a.	5490	294	1.9	1.81	
Toray	Torayca	T800HB	12	445	n.a.	5490	294	1.9	1.81	
Toray	Torayca	T830HB	6	223	n.a.	5340	294	1.8	1.81	
Toray	Torayca	T1000GB	12	485	n.a.	6370	294	2.2	1.80	
Toray	Torayca	T1100GC	12	505	n.a.	7000	324	2.0	1.79	
Toray	Torayca	T1100GC	24	1010	n.a.	7000	324	2.0	1.79	

Manufacturer	Brand Name	Fiber Name	Number of Filaments in K	Linear Density in tex	Diameter in µm	Tensile Strength in MPa	Tensile Modulus in GPa	Elongation in %	Density in g/cm³	Fiber Type/ Remark
Toray	Torayca	M35JB	6	225	n.a.	4510	343	1.3	1.75	CAR-BIDE
Toray	Torayca	M35JB	12	450	n.a.	4700	343	1.4	1.75	
Toray	Torayca	M40JB	6	225	n.a.	4400	377	1.2	1.77	
Toray	Torayca	M40JB	12	450	n.a.	4400	377	1.2	1.77	
Toray	Torayca	M46JB	6	223	n.a.	4200	436	1.0	1.84	
Toray	Torayca	M46JB	12	445	n.a.	4020	436	0.9	1.84	
Toray	Torayca	M50JB	6	216	n.a.	4120	475	0.9	1.88	
Toray	Torayca	M55J	6	218	n.a.	4020	540	0.8	1.91	
Toray	Torayca	M55JB	6	218	n.a.	4020	540	0.8	1.91	
Toray	Torayca	M60JB	3	103	n.a.	3820	588	0.7	1.93	
Toray	Torayca	M60JB	6	206	n.a.	3820	588	0.7	1.93	
Toray	Torayca	M30SC	18	760	n.a.	5490	294	1.9	1.73	
Solvay Cytec	Thornel PAN	T300	1	66	7.0	3750	231	1.6	1.76	HT
Solvay Cytec	Thornel PAN	T300	3	198	7.0	3750	231	1.6	1.76	
Solvay Cytec	Thornel PAN	T300	6	398	7.0	3750	231	1.6	1.76	
Solvay Cytec	Thornel PAN	T300	12	792	7.0	3750	231	1.6	1.76	
Solvay Cytec	Thornel PAN	T650	3	196	6.8	4280	255	1.7	1.77	
Solvay Cytec	Thornel PAN	T650	6	395	6.8	4280	255	1.7	1.77	
Solvay Cytec	Thornel PAN	T650	12	777	6.8	4280	255	1.7	1.77	
Solvay Cytec	Thornel Pitch	P25	2	320	10.0	1560	159	0.9	1.92	pitch-based
Solvay Cytec	Thornel Pitch	P25	4	715	11.0	1560	172	0.9	1.90	
Solvay Cytec	Thornel Pitch	P30	2	320	10.0	1380	207	0.8	2.00	
Solvay Cytec	Thornel Pitch	P55	4	714	10.0	1380	414	0.5	2.00	

Table 3.13 Supplementary Table to Figure 3.29: Manufacturer, Brand Name, Fiber Designation, and Properties *(continued)*

Manufacturer	Brand Name	Fiber Name	Number of Filaments in K	Linear Density in tex	Diameter in µm	Tensile Strength in MPa	Tensile Modulus in GPa	Elongation in %	Density in g/cm³	Fiber Type/ Remark
Mitsubishi Chemical Corp.	PYROFIL	TR30S	3	200	7.0	4410	235	1.9	1.79	HT
Mitsubishi Chemical Corp.	PYROFIL	TR50S	6	400	7.0	4900	235	2.1	1.82	
Mitsubishi Chemical Corp.	PYROFIL	TR50S	12	800	7.0	4900	235	2.1	1.82	
Mitsubishi Chemical Corp.	PYROFIL	TR50S	15	1000	7.0	4900	235	2.1	1.82	
Mitsubishi Chemical Corp.	PYROFIL	TR50D	12	800	7.0	5000	235	2.1	1.82	
Mitsubishi Chemical Corp.	PYROFIL	TRH50	12	800	7.0	4900	250	2.0	1.82	
Mitsubishi Chemical Corp.	PYROFIL	TRH50	18	1000	6.0	5300	250	2.1	1.82	
Mitsubishi Chemical Corp.	PYROFIL	TRH50	60	3200	6.0	4830	250	1.9	1.81	
Mitsubishi Chemical Corp.	PYROFIL	TRW40	50	3750	7.0	4120	240	1.7	1.81	
Mitsubishi Chemical Corp.	PYROFIL	MR60H	24	960	5.0	5680	280	2.0	1.81	IM
Mitsubishi Chemical Corp.	PYROFIL	MR70	12	415	5.0	7000	324	2.2	1.82	

Manufacturer	Brand Name	Fiber Name	Number of Filaments in K	Linear Density in tex	Diameter in µm	Tensile Strength in MPa	Tensile Modulus in GPa	Elongation in %	Density in g/cm³	Fiber Type/ Remark
Mitsubishi Chemical Corp.	PYROFIL	MS40	12	600	6.0	4410	340	1.3	1.77	CAR-BIDE
Mitsubishi Chemical Corp.	PYROFIL	HR40	12	600	6.0	4410	375	1.2	1.82	
Mitsubishi Chemical Corp.	PYROFIL	HS40	12	430	5.0	4610	425	1.1	1.85	
Mitsubishi Chemical Corp.	Grafil	34-700	12	800	7.0	4830	234	2.1	1.80	HT
Mitsubishi Chemical Corp.	Grafil	34-700	24	1600	7.0	4830	234	2.1	1.80	
Mitsubishi Chemical Corp.	Grafil	34-800	30	1675	6.0	5520	255	2.2	1.81	
Mitsubishi Chemical	DIALEAD	K1352U	2K, 6K	270	n.a.	3600	620	0.6	2.12	pitch-based
Mitsubishi Chemical	DIALEAD	K1392U	2K, 6K	270	n.a.	3700	760	0.5	2.15	
Mitsubishi Chemical	DIALEAD	K13C2U	2K, 6K	270	n.a.	3800	900	0.4	2.20	
Mitsubishi Chemical	DIALEAD	K13C6U	2K, 6K	760	n.a.	3600	900	0.4	2.18	
Mitsubishi Chemical	DIALEAD	K13D2U	2K, 6K	365	n.a.	3700	935	0.4	2.20	
Mitsubishi Chemical	DIALEAD	K13312	12K, 16K	1560	n.a.	3200	420	0.7	2.06	
Mitsubishi Chemical	DIALEAD	K63712	12K, 16K	2000	n.a.	2600	640	0.4	2.12	
Mitsubishi Chemical	DIALEAD	K13916	12K, 16K	2200	n.a.	3200	760	0.4	2.15	
Mitsubishi Chemical	DIALEAD	K63A12	12K, 16K	1950	n.a.	2600	790	0.3	2.12	

Table 3.13 Supplementary Table to Figure 3.29: Manufacturer, Brand Name, Fiber Designation, and Properties *(continued)*

Manufacturer	Brand Name	Fiber Name	Number of Filaments in K	Linear Density in tex	Diameter in µm	Tensile Strength in MPa	Tensile Modulus in GPa	Elongation in %	Density in g/cm³	Fiber Type/ Remark
UMATEX	UMATEX	UMT40	3K	180	n.a.	4000	260	1.5	1.77	HT
UMATEX	UMATEX	UMT45	12K	790	n.a.	4500	260	1.7	1.78	
UMATEX	UMATEX	UMT45	24K	1600	n.a.	4500	260	1.7	1.78	
UMATEX	UMATEX	UMT49	12K	770	n.a.	4900	260	1.8	1.78	
UMATEX	UMATEX	UMT400	12K	720	n.a.	4500	400	1.0	1.82	IM
UMATEX	UMATEX	UMT430	12K	700	n.a.	4500	430	0.9	1.84	
UMATEX	UMATEX	UMT450	12K	680	n.a.	4500	450	0.9	1.85	
Hyosung	HT	H2550	6	400	7.0	5516	250	2.2	1.80	HT
Hyosung	HT	H2550	12	800	7.0	5516	250	2.2	1.80	
Hyosung	HT	H2550	24	1650	7.0	5516	250	2.2	1.80	
Hyosung	IM	H3055	12	725	6.6	5516	290	1.9	1.80	IM
SGL Carbon	SIGRAFIL	CT24-4.8	24	1600	7.0	4800	240	2.0	1.81	HT
SGL Carbon	SIGRAFIL	CT24-5.0	24	1600	7.0	5000	270	1.9	1.81	
SGL Carbon	SIGRAFIL	CT50-4.0	50	3300	6.8	4000	240	1.7	1.80	
SGL Carbon	SIGRAFIL	CT50-4.4	50	3420	7.0	4400	255	1.65	1.78	
SGL Carbon	SIGRAFIL	CT50-4.8	50	3070	6.6	4800	280	1.65	1.78	AM
Formosa Plastics	Tairyfil	TC-33	1.5	100	7.0	3450	230	1.5	1.80	HT
Formosa Plastics	Tairyfil	TC-33	3	200	7.0	3450	230	1.5	1.80	
Formosa Plastics	Tairyfil	TC-33	6	400	7.0	3450	230	1.5	1.80	
Formosa Plastics	Tairyfil	TC-35	3	200	7.0	4000	240	1.6	1.80	
Formosa Plastics	Tairyfil	TC-35	6	400	7.0	4000	240	1.6	1.80	
Formosa Plastics	Tairyfil	TC-35	12	800	7.0	4000	240	1.6	1.80	
Formosa Plastics	Tairyfil	TC-35	18	1200	7.0	4000	240	1.6	1.80	

Manufacturer	Brand Name	Fiber Name	Number of Filaments in K	Linear Density in tex	Diameter in µm	Tensile Strength in MPa	Tensile Modulus in GPa	Elongation in %	Density in g/cm³	Fiber Type/ Remark
Formosa Plastics	Tairyfil	TC-35	24	1600	7.0	4000	240	1.6	1.80	HT
Formosa Plastics	Tairyfil	TC-35	48	3200	7.0	4000	240	1.6	1.80	
Formosa Plastics	Tairyfil	TC-36S	12	800	7.0	4900	250	2.0	1.81	
Formosa Plastics	Tairyfil	TC-36S	24	1600	7.0	4900	250	2.0	1.81	
Formosa Plastics	Tairyfil	TC-42S	12	440	5.1	5690	290	2.0	1.81	
Formosa Plastics	Tairyfil	TC-42S	24	880	5.1	5690	290	2.0	1.81	
Formosa Plastics	Tairyfil	TC-55	12	420	5.0	4400	380	1.2	1.79	IM
Formosa Plastics	Tairyfil	TC-55	24	840	5.0	4400	380	1.2	1.79	
Dow Aksa	Dow Aksa	A-38	3	200	n.a.	3800	240	1.6	1.78	HT
Dow Aksa	Dow Aksa	A-38	6	400	n.a.	3800	240	1.6	1.78	
Dow Aksa	Dow Aksa	A-42	12	800	n.a.	4200	240	1.8	1.78	
Dow Aksa	Dow Aksa	A-42	24	1600	n.a.	4200	240	1.8	1.78	
Dow Aksa	Dow Aksa	A-45	12	800	n.a.	4500	240	1.9	1.79	
Dow Aksa	Dow Aksa	A-49 D	12	800	n.a.	4900	250	2.0	1.79	
Dow Aksa	Dow Aksa	A-49 D	24	1600	n.a.	4900	250	2.0	1.79	
Zoltek	Panex	PX35	50	3300	7.2	4137	242	1.6	1.81	HT

4 Characterization of Carbon Fibers

This chapter presents methods for testing the properties of carbon fibers. Methods that are essential for the development of new carbon fibers will also be discussed.

The properties of fiber composite plastics (fiber-reinforced plastics, FRP) depend essentially on the fiber used and its properties as well as on the matrix polymer used. Therefore, it is of decisive importance for the application of FRP that the fibers used always have constant properties. In order to ensure this, the fibers must be tested for constant material properties in the sense of quality assurance before they are processed into composite materials. Within a fiber batch, related to the batch size/run length, a confidence level of 95% is usually required to be able to speak of constant properties. The 95% confidence range is defined as the mean value of a measured variable and is given with the same accuracy as the mean value itself (see also ISO 1973).

Carbon fiber characterization tests are carried out either on single filaments or on rovings. The anisotropic material behavior of carbon fibers in comparison with, for example, metallic materials makes special test procedures necessary. Since the production of carbon fibers as continuous yarn is a continuous process (see Chapter 3), but there are no online test methods for continuous fiber testing in the process, the properties of the yarn can always only be determined on the basis of samples at specified intervals. Today there are some (partially) automated testing devices that considerably simplify the performance of tests while ensuring very good reproducibility of the measurements. Typical standards and parameters essential for fiber production are summarized in Table 4.1.

Table 4.1 Overview of Relevant Standards for the Characterization of Yarn Properties in Fiber Production

Feature	Tensile Strength	Tension Modulus	Fracture Elongation	Linear Density	Sizing Level	Density	Filament Diameter
	MPa	GPa		tex		g/cm³	µm
Standard	ISO 10618	ISO 10618	ISO 10618	ISO 1889	ISO 10548	ISO 10119	ISO 1973
Typical repetition frequency	4 to 8 spools every 6 to 8 h = 12 to 32 spools/day					1 coil/day	1 coil/day

In addition to the pure material properties, from the processor's point of view, processing-relevant properties of the dry fiber material, such as spreadability and friction behavior of the yarn, can also be tested. Even if the fiber properties listed in Table 4.1 are subject to only a small variation from production batch to production batch, there are differences which are reflected in the processing behavior of different fiber batches. This can manifest itself in the manufacture of fabrics, for example, shown by the fact that the fibers are difficult to weave and that fiber breakage and lint (fuzz) formation occur. An important parameter for the application of FRP is the fiber-matrix connection (FMA), which describes the interaction of fiber and matrix in the composite (see also Chapter 3). A number of non-standard processes have been developed to classify the processing properties of carbon fiber yarns (Table 4.2).

Table 4.2 Test Methods for Characterizing and Classifying the Processing Behavior of Carbon Fiber Yarns and Filaments

Property/Error	Test Method/Procedure	Application Reference
Drapability/Formability	Drape test	Forming of fabrics and scrims for complex component geometries
Impregnability	Contact angle measurement	Fiber wetting or impregnability of fabrics and scrims with matrix
Fiber breakage, tendency to lint formation (fuzz level)	Roving test	Manufacture of fabrics and scrims; fiber breakage increases, as a rule, with increasing modulus of elasticity
Yarn spreading (gaps in the fiber bed)		Production of fabrics and scrims, influence on impregnation behavior
Twists in the yarn bundle (gaps in the fiber bed)		Production of UD tapes and fabrics

This chapter does not deal with other test procedures relevant for carbon fiber-based composites such as destructive or non-destructive testing or (thermo-) mechanical testing. In this context, the reference to already existing technical literature should suffice [4-1] [4-2].

4.1 Test Methods on Dry Yarn

A large number of fiber properties can already be characterized on dry yarn. In the following, methods for the determination of chemical-physical properties such as the density, the filament diameter or the tensile strength of a filament are briefly described. In addition, processes will be discussed that focus on the macroscopic processing properties of carbon fibers.

4.1.1 Density Determination

The density determination of carbon fibers is used to check the degree of carbonization, which significantly determines the mechanical properties of the fiber. The density of a carbon fiber can be determined by three different methods:

- Method A: Liquid displacement (Archimedean principle)
- Method B: Swim-sink method
- Method C: Density gradient column (reference method)

In preparation for the density determination, the sample is freed from the preparation application (sizing) before measurement. Methods such as solvent extraction, pyrolysis, or chemical digestion can be used to remove the sizing. All samples are conditioned at standard climate.

Method A (Displacement Method)

The method is based on the Archimedean principle. To determine the density, a fiber sample is cut out of a yarn bobbin and formed into a knot or bundle so that the sample fits into the measuring apparatus. First, the dry mass of the fiber sample is determined by weighing. The sample is then completely immersed in a suitable test liquid, the density of which has been determined precisely beforehand using a pycnometer. The density of the fiber ρ_Θ in g/cm^3 is given by Equation 4.1:

$$\rho = \frac{w_1}{w_1 - w_2} * \rho_L \tag{4.1}$$

w_1 = sample weight in air in g
w_2 = sample weight in test liquid in g
ρ_L = density of the test liquid in g/cm^3

Method B (Swim-Sink Method)

The method is based on the principle that the fiber sample is immersed in a mixture of two liquids whose density is changed until it corresponds to the fiber density. This procedure also follows the Archimedean principle. First, a mixture of two different liquids (each with a different density) is produced in a mixing bottle in such a way that the density of the mixture is lower than that of the carbon fiber. A sample of the fiber is dipped into the mixture in the form of a knot. The liquid with a higher density is then added to the mixture drop by drop until the fiber sample finally floats in the liquid mixture and no longer sinks. The fiber sample is then removed and the density of the liquid mixture is determined using a pycnometer. The density ρ of the tested carbon fiber in g/cm^3 corresponds to the result of the pycnometric measurement.

Method C (Density Gradient Column)

This method is based on the principle that a sample is immersed in a liquid column with a linear density gradient and does not sink or rise further within the liquid column according to the sample density. *Density gradient columns* are columns whose density increases uniformly from top to bottom. Thus, method C is basically based on the same principle as method B. First, a gradient column (height 1 m, ⌀ 40 to 50 mm) is produced from two liquids of different density. The density distribution in the column is determined using several calibrated reference bodies in the range of the expected carbon fiber sample density and plotted as a calibration curve. A quantity of 1 to 10 mg of the carbon fiber is dipped into the mixture from above in the form of a knot or bundle. The sample slowly sinks down in the column until an equilibrium is reached at the same density of fiber sample and liquid mixture and the fiber sample floats in the liquid. The density ρ of the sample in g/cm^3 results from the previously determined calibration curve of the gradient column.

Normative References:

- ISO 10119:2002: Carbon fibre – Determination of density
- ISO 10548:2002: Carbon fibre – Determination of size content
- ISO 291: Plastics – Standard atmospheres for conditioning and testing

4.1.2 Determination of Linear Density

Gravimetric Determination

The principle for determining the linear density is based on the gravimetric determination of a fiber bundle sample of defined and known length. The measurement can be carried out on carbon fiber yarn with or without preparation application (sizing). The sample is conditioned or dried under a standard climate before measurement. The sizing can be removed by solvent extraction, pyrolysis or chemical digestion. In this case, drying of the sample is not required.

At first the empty weight of a suitable bobbin is measured. Then it is wounded with a damage-free fiber sample of defined length. According to Table 4.3 sample length depends on the nominal linear density of the fiber.

Table 4.3 Overview Sample Lengths vs. Nominal Linear Density for Carbon Fibers

Nominal Linear Density T_f in tex	Sample Length in m
$T_f < 50$	Select length so that mass ≥ 0.25 g
$50 < T_f < 125$	5
$125 < T_f < 250$	2
$250 < T_f$	1

The mass m of the fiber sample is calculated from the difference between the two weighs (weight of empty spool or wound spool).

The linear density T_f in g/km of the fiber is obtained from the following equation and is expressed in tex:

$$T_f = \frac{1000 * m}{L}$$

m = mass of the sample in g
L = sample length in m

Vibration Method

Another method for determining linear density is based on the principle of vibroscopic measurement (ISO 1973). Figure 4.1 shows a machine developed by *Textechno GmbH* which largely automates the measurement [4-3]. For this purpose, a filament from a bundle of yarns is inserted into the measuring apparatus and vibrated by a sinusoidal frequency sweep via a loudspeaker. The linear density can be calculated from the resonant frequency f of the filament measured at a constant measuring length L and a known preload F_V according to Equation 4.2:

$$T_f = \frac{F_V}{4 * f^2 * L^2} \tag{4.2}$$

T_f = linear density
F_V = preload force
f = resonant frequency
L = free sample length

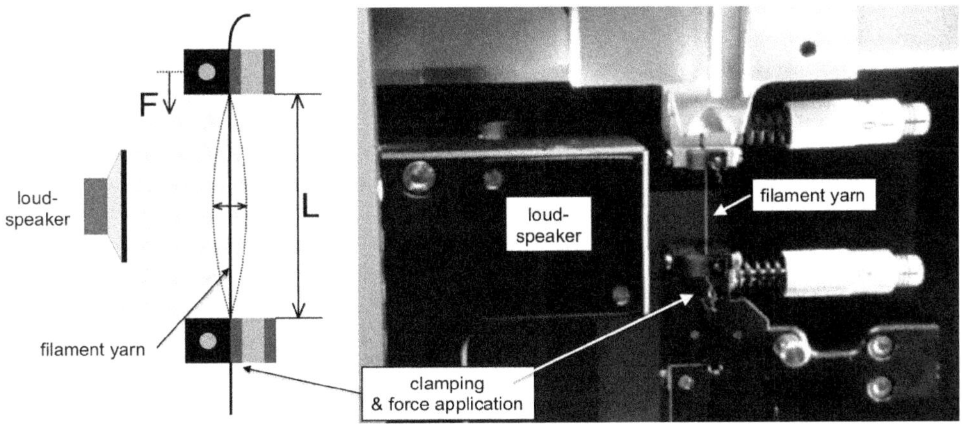

Figure 4.1 Measuring arrangement for vibroscopic determination of linear density [*Textechno GmbH*]

Normative References:

- ISO 1889: Reinforcement yarns – Determination of linear density
- ISO 1973:1995-12: Textile fibres – Determination of linear density – Gravimetric method and vibroscope method
- ASTM D 1577: Standard Test Methods for Linear Density of Textile Fibers
- ISO 291: Plastics – Standard atmospheres for conditioning and testing
- ISO 10548:2002: Carbon fibre – Determination of size content

4.1.3 Determination of the Filament Diameter

The filament diameter can be determined in accordance with ISO 1973 using the same vibroscopic measuring arrangement (Figure 4.11) as that used to determine the linear density. The filament cross-sectional area A can be calculated from the filament at constant measuring length L and known pretension F_V as well as the resonant frequency f of the filament according to the equation below. In the case of a circular cross-section, the filament diameter also results as follows:

$$A = \frac{F_V}{4 * \rho * f^2 * L^2} = 2 * \sqrt{\frac{A}{\pi}} \tag{4.3}$$

A = filament cross-sectional area
d = filament diameter
ρ = density of the fiber

Normative References:

- ISO 1973:1995-12: Textile fibres – Determination of linear density – Gravimetric method and vibroscope method
- ASTM D 1577: Standard Test Methods for Linear Density of Textile Fibers

4.1.4 Determination of the Preparation Mass Fraction

Preparations or sizings of carbon fibers (see also Chapter 3) are all types of materials that are applied to the fibers to improve handling and processing. The proportion of preparation applied to the surface of carbon fiber yarns during fiber manufacture indicates the proportion of preparation (sizing) to the total mass of the yarn (fiber bundle) and can be determined in accordance with ISO 10548. It is defined as the percentage ratio of the mass change Δm of a fiber sample by removing the sizing relative to the total mass m of the sample.

To determine this percentage of the preparation mass (sizing level), three methods are mentioned in the standard:

- **Method A:** Removal of the preparation by Soxhlet extraction; used when the sizing is completely soluble in a suitable solvent.
- **Method B:** Removal of the preparation by wet chemical oxidation with a sulfuric acid/hydrogen peroxide mixture; applied when the preparation is partially cured and completely soluble in solvents.
- **Method C:** Removal of preparation by high temperature pyrolysis in a nitrogen atmosphere; used when preparation is completely removed by pyrolysis.

All three methods are based on the principle that the preparation mass fraction is determined gravimetrically by differential weight measurement. A fiber sample is first weighed and then the preparation is removed according to one of the three methods described above. Finally, the sample is weighed again. The proportion of preparation mass results accordingly from the difference in weight Δm. The result is the mass fraction of the sizing SC (size content) in percent.

$$\mathrm{SC} = \frac{\Delta m}{m} * 100\% \tag{4.4}$$

Δm = difference in weight
m = mass of sample at initial sample weight

Normative References:

- ISO 10548:2002: Carbon fibre - Determination of size content

4.1.5 Determination of Mechanical Parameters on Dry Yarn

The determination of mechanical parameters such as tensile strength, modulus of elasticity, and elongation at break on dry yarn can be carried out on individual filaments. Due to the fineness of individual filaments, this requires special attention during test specimen preparation as well as a sufficiently large number of specimens (at least 20 samples) in order to obtain meaningful characteristic values. As with other test methods mentioned above, the sizing can be removed beforehand (not absolutely necessary). In addition to the tensile test, a measurement of the filament diameter independent of this method (see Section 4.1.3) is required for the correct calculation of all mechanical characteristic values. According to ISO 11566, a carbon fiber single filament is fixed under low pre-tension with a suitable adhesive in the specimen holder at both ends and loaded in the tensile testing machine at a test speed of 1 to 5 m/min to break. The tensile strength is determined by the following:

$$\sigma_f = \frac{F}{A} \tag{4.5}$$

F = maximum tractive force in N
A = cross-sectional area of the single filament in mm^2

Based on this procedure, adapted testing machines have been developed and are commercially available (e.g. FAVIMAT+, from *Textechno GmbH*), which, in addition to a significantly simplified and automated specimen collection of the individual filaments, also determines the filament diameter of the specimen before the tensile test according to ISO 1973 (vibroscopic measurement; see Section 4.1.3) [4-4]. Time-consuming and cumbersome determination of the sample diameter using microscopic methods on the basis of CFRP specimens is no longer necessary. Today, up to 500 samples can be tested automatically in series. As a rule, about 50 samples of the same fiber type are sufficient to achieve a high reproducibility.

Normative References:

- ISO 11566: Carbon fibre – Determination of the tensile properties of single-filament specimens
- ISO 10618: Carbon fibre – Determination of tensile properties of resin-impregnated yarn

4.1.6 Roving Test

The roving test is a test method which has not yet been standardized and which enables the testing of various processing-relevant fiber properties [4-5]. The method represents a modular combination of several different test modules and allows the automated characterization of different processing-relevant fiber properties. The following properties of the yarn can be characterized:

- Friction properties: friction within the roving (fiber-to-fiber) as well as friction behavior against solid surfaces (abrasion) and determination of the friction coefficient.
- Spreading behavior: how light and how strong can the fiber bundle be spread evenly without gaps between the filaments?
- Detection of broken filaments: before/after the spreading process or friction on a surface.
- Testing for fiber breakage and fuzz (fiber fluff) formation; this increases with increasing modulus of elasticity.

- Homogeneity of the filament orientation in the yarn bundle: Detection by means of eddy current testing of the extent to which the filaments in the yarn are not perfectly oriented.
- Twists in the yarn ("false twist"); leads to e.g. gaps or weaving flaws.
- Control of the winding angle; it differentiates between e.g. HT and IM fiber yarns.

Figure 4.2 shows a test stand with the individual modules and description.

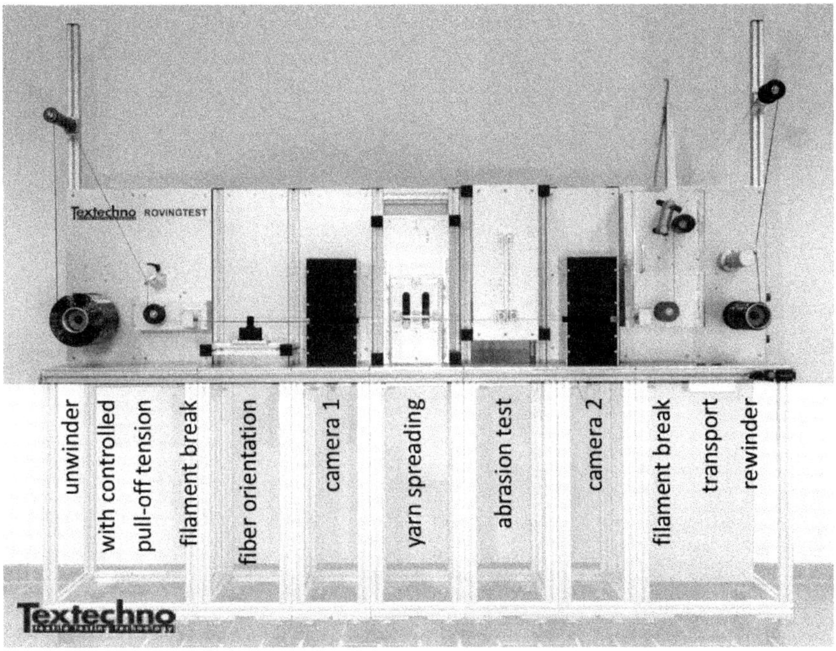

Figure 4.2 Test bench: roving tester [*Textechno GmbH*]

4.1.7 Drape Test

The drape test is used to characterize the drapability of technical textiles for use in fiber composite plastics (FRP). In the automated production of FRP, undetected defects in formed fabrics and scrims such as gaps, undulations, folds, and waves, which typically occur in three-dimensional forming, represent a quality problem (Figure 4.3).

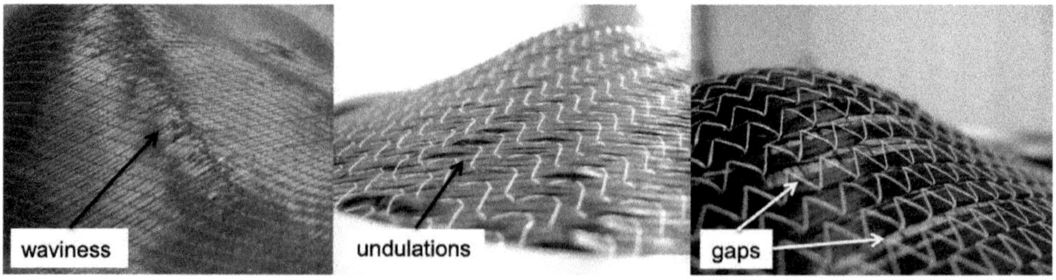

Figure 4.3 Examples of drapery errors due to forming [*Textechno GmbH*]

Such defects remain after impregnation of the textile with a matrix resin and subsequent curing, which can lead to a weakening of the finished component. In the procedure described in DIN SPEC 8100, carbon fiber fabrics are deformed with the aid of a reference body. The formation of possible imperfections during draping is detected by a high-resolution camera in conjunction with image analysis. The characteristic values determined by this method can be used to characterize the draping behavior of fabrics and scrims. The procedure can thus be used to validate the simulative description and calculation in the design of FRP components.

The test is performed on a round textile blank, which is deformed in the test device by means of a calotte-shaped reference body. For this purpose, the blank is clamped around its edge in a blank holder, the clamping force of which is controlled and checked by the testing software. At the same time, the forces applied by the reference body during forming are measured (Figure 4.4, steps 1–3). A high-resolution camera is then used to take pictures of the deformed textile cut and analyze them for gaps, fiber angle deviations, etc. (Figure 4.4, step 4). As a result, percentage diagrams showing, for example, the gap width distribution or the deviation from a given fiber angle as a function of the penetration depth of the reference body into the textile are generated (Figure 4.5).

Normative References:

- DIN SPEC 8100: Reinforcement textiles – Automatic measurement of drapability on non-crimped and woven fabrics for continuous fiber reinforced materials

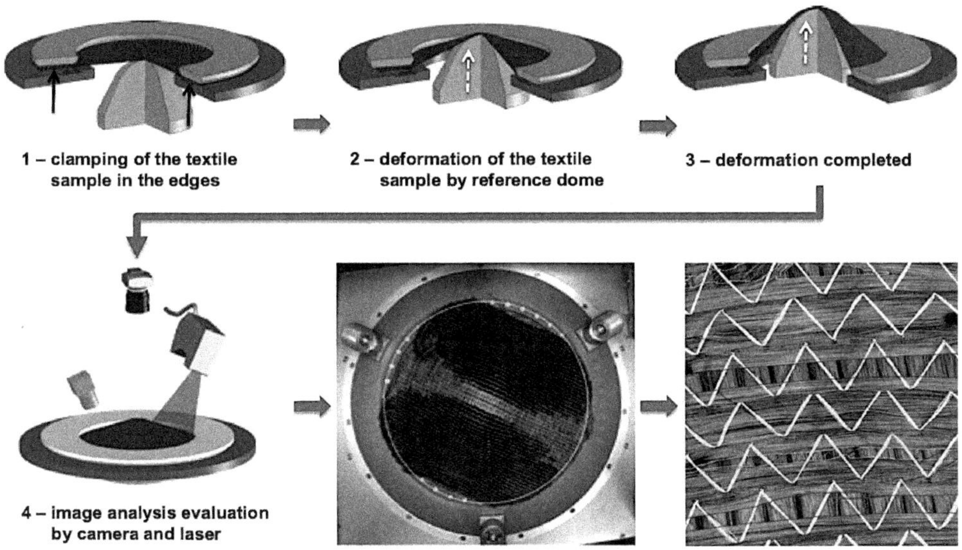

Figure 4.4 Schematic sequence of the drape test [*Textechno GmbH*]

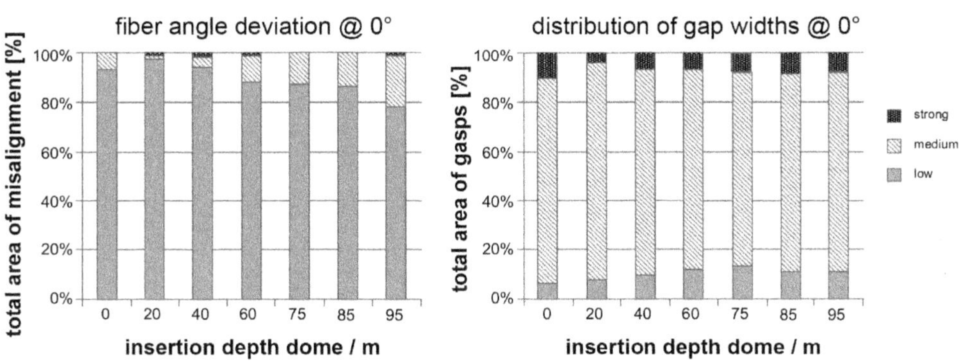

Figure 4.5 Examples of results from the drape test [*Textechno GmbH*]

4.1.8 Surface Activity and Impregnability

The surface activity and thus the impregnation capability of carbon fiber yarns play an important role in the processing of the fiber with a liquid matrix to a fiber composite material. The impregnation ability of yarns can be determined by the method of contact angle measurement. The contact angle or wetting angle is the angle that a drop of liquid on the surface of a solid forms to this surface (Figure 4.6, left).

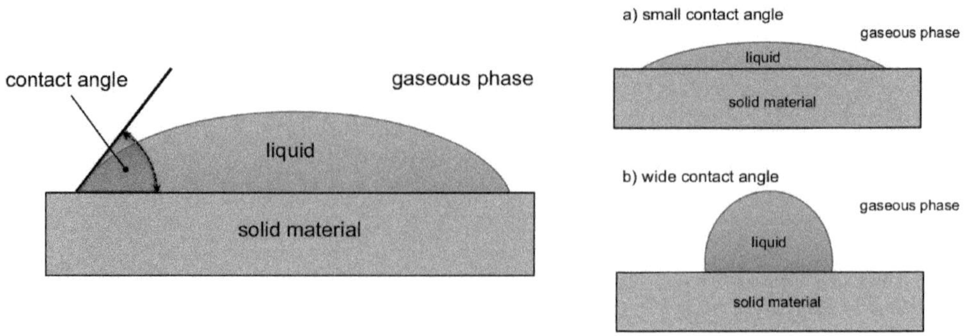

Figure 4.6 Left: principle of determining the contact angle; right: examples of contact angles

The size of the contact angle between liquid and solid depends on interactions at the contact surface between the substances. From the determination of the contact angle, properties of the fiber surface of the yarn, e.g. the surface energy, can be determined. The latter determines how well a carbon fiber yarn can be wetted and impregnated. This means that the better the surface energies of the yarn and the matrix match, the lower the contact angle and the better the impregnation capacity of the yarn (Figure 4.6, top right). By appropriate treatment of the yarn surface, e.g. with a preparation application (sizing), the wetting behavior and thus the contact angle can be changed. The contact angle can be measured with a contact angle goniometer.

4.2 Test Method on Impregnated Yarn

In addition to the pure fiber properties, the properties of the composite of fiber and matrix are particularly important in terms of application. If these two partners do not fit together optimally, the fiber composite material fails prematurely or does not even reach a certain property level.

4.2.1 Determination of Tensile Strength

While the determination of the tensile strength on a dry single filament (see Section 4.1.5) requires a large number of specimens (at least 20) in order to be able to make a reliable statement on the tensile strength of a yarn bundle, the tensile test in accordance with ISO 10618 or DIN 65382, a meaningful result can already be determined with three valid measurements. Furthermore, the samples tested here can be used for microscopic examination (SEM, see Section 4.2.2.5) of the fiber-matrix connection. Due to today's automation of single filament tests, even large numbers of samples can be tested in a short time (approx. 1 minute per test).

For the production of the test specimen, a yarn sample is evenly impregnated with a reaction resin and then cured. Alternatively, thermoplastic matrix systems can also be used. It is essential for the production that the impregnating resin is compatible with the yarn and its preparation and has a sufficiently low viscosity for complete and uniform impregnation of the carbon yarn, so that a sufficient fiber volume fraction (UD fibers approx. 60%, fabric 50%) is achieved in the composite. The impregnation of the yarn bundle with the polymer matrix improves the inner bond between the filaments, so that the load introduction can be evenly distributed to all filaments in the bundle. Samples with different force introduction elements can be used for load introduction (see Figure 4.7).

The elongation at break of the matrix used must be at least twice and preferably three times as large as that of the fiber. The hardened sample should not contain any broken or wrongly oriented filaments or resin accumulations.

The specimens are subjected to tensile loading in a mechanical testing machine at constant speed until fracture (see also ISO 527-1 and ISO 527-4). The tensile strength σ_f in megapascals (MPa) and the modulus of elasticity E_f in gigapascals (GPa) as well as the elongation ε_E in percent (%) can be calculated from the measured force-length change. The exact procedure is described in detail in the relevant literature [4–7].

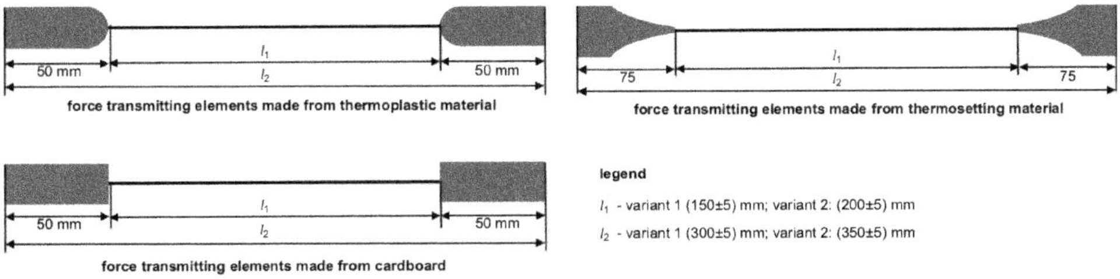

Figure 4.7 Force introduction elements for determining tensile properties

Normative References:

- ISO 10618: Carbon fibre - Determination of tensile properties of resin-impregnated yarn
- DIN 65382: Aerospace; reinforcement fibres for plastics; tensile test of impregnated yarn test specimens
- ISO 527-1: Plastics - Determination of tensile properties - Part 1: General principles
- ISO 527-4: Plastics - Determination of tensile properties - Part 4: Test conditions for isotropic plastics and anisotropic fiber-reinforced plastic composites

4.2.2 Fiber-Matrix Interphase

The FMA (or fiber-matrix adhesion) is a very sensitive indicator for checking the compatibility between fiber and matrix. This binding can be sensitively influenced by parameters such as the surface activity of the carbon fiber and the preparation application (type, quantity) (see also Chapter 3). The performance of a composite material depends largely on the FMA.

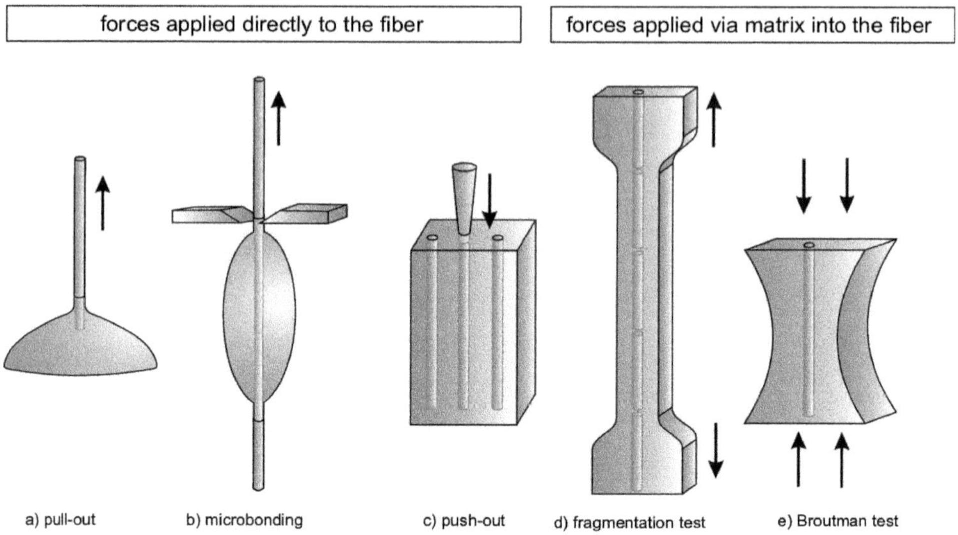

Figure 4.8 Micromechanical test methods and their force introduction [4-6]

A number of different micromechanical tests are available to determine this interfacial behavior. These are shown schematically in Figure 4.8 and can be divided into two groups on the basis of the force introductions:

- Force applied directly to the fiber
- Force introduction via the matrix into the fiber

In all of the methods described above, sample preparation plays an important role, as the measurements are very sensitive even to small disturbances. In fact, all of the methods shown above show a wide dispersion, so that the major carbon fiber manufacturers have mostly developed their own tests. In addition to the wide dispersion of the measurement results from individual filament tests, the transferability to yarn bundles must also be considered with caution, since the properties of the yarn bundles are relevant for the subsequent CFRP component. However, the test methods described above can provide valuable information for the observation and development of fiber preparations (sizing). Some of the test procedures described are discussed below.

4.2.2.1 Testing by Pull-Out Test

For testing the fiber-matrix adhesion by pull-out test, a single filament is embedded vertically in a matrix drop by means of a suitable testing device (Figure 4.9, right) [4–8]. The matrix can be either thermoplastic or reaction systems. It is important that the fiber is absolutely perpendicular in the matrix so that the subsequent tensile test does not result in transverse forces in the fiber that falsify the result. Under tensile load, the filament is slowly stretched until the first failure of the fiber-matrix bond occurs, which is shown in the tensile-path diagram (Figure 4.9, left) as a slight kink in the curve. This point is marked as F_d (debonding force). When the maximum force F_{max} is reached, the fiber-matrix adhesion fails completely and the fiber begins to slip out of the matrix. A rapid decrease in force occurs until only frictional forces (F_b) exist between the binding partners and the fiber slides further out of the matrix. The embedding length l_e of the fiber in the matrix can be read from the total distance travelled. The following additional parameters can be calculated from the results:

Apparent interfacial shear strength	τ_{app} in MPa
Local interface shear strength	τ_d in MPa
Critical interfacial energy release rate	G_{Ic} in J/m^2

The apparent interfacial shear strength τ_{app} is based on the achieved maximum force F_{max} and is suitable as a first indicator for a relative comparison of the fiber-matrix adhesion. The result is influenced by the interfacial friction of the binding partners. The local interfacial shear strength τ_d is based on the release force F_d (debonding force) and is an absolute measure of the fiber-matrix adhesion independent of the interfacial friction, corrected for the deformation of the fiber and matrix during the pull-out test. The critical interfacial energy release rate G_{Ic} describes the energy required to detach the fiber surface from the matrix.

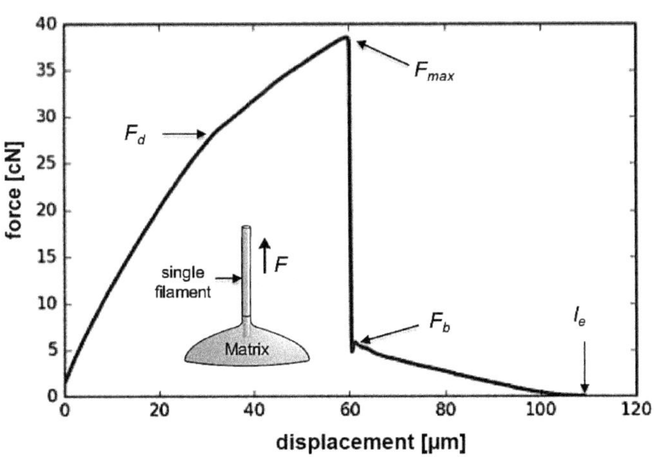

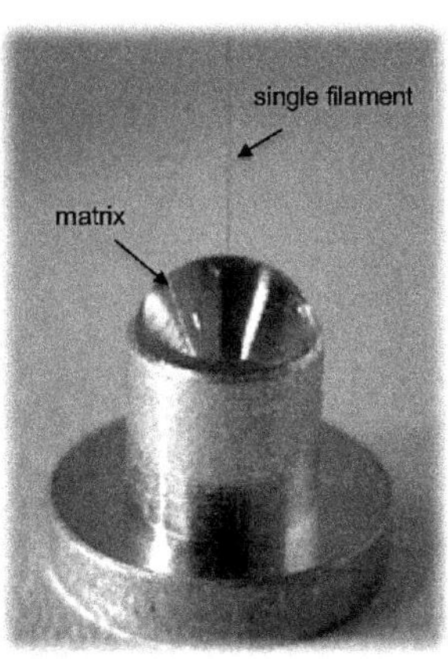

Figure 4.9 Left: result diagram pull-out tests with the fiber-matrix adhesion tester; right: filament embedded in matrix drops [*Textechno GmbH*]

Figure 4.10 shows examples of force-displacement diagrams for determining the FMA of a carbon fiber without or with an adapted preparation application in polyamide. It can be seen that the measurements sometimes show large fluctuations in the course of the force. The force increase of fibers with adapted sizing can be clearly seen in the pull-out test, which indicates an improved fiber-matrix connection.

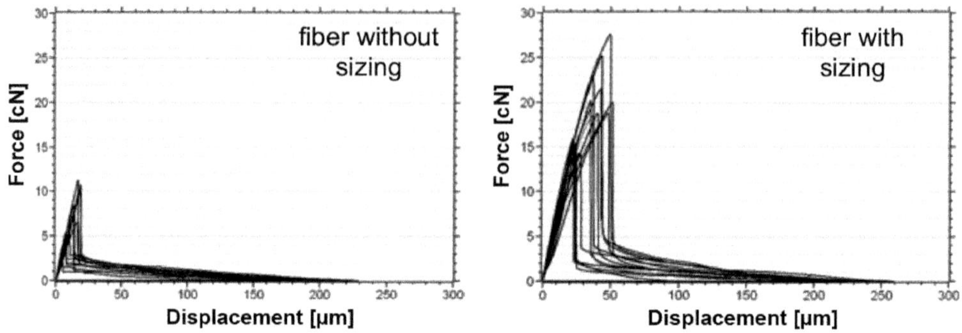

Figure 4.10 Force-displacement diagrams for determining the FMA of a carbon fiber in polyamide with different preparation application [*Textechno GmbH*]

The influence of the preparation application on the fiber can be seen even more clearly in Figure 4.11. The three parameters apparent interfacial shear strength τ_{app}, local interfacial shear strength τ_d, and energy release rate G_{Ic} clearly show the positive influence of sizing.

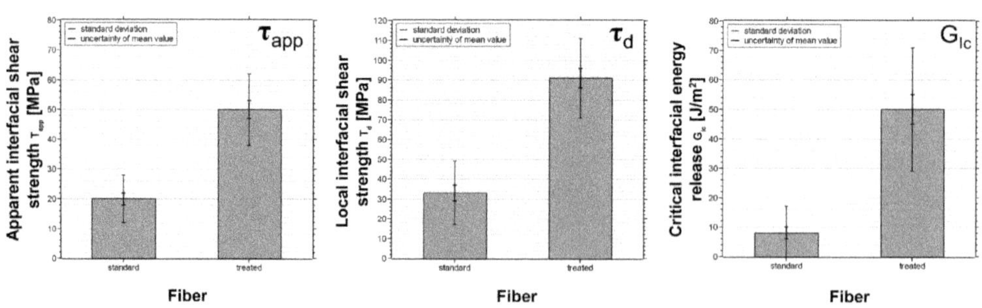

Figure 4.11 Comparison of the FMA results of a carbon fiber with different preparation application in polyamide [*Textechno GmbH*]

4.2.2.2 Testing by Push-Out Test

Another possibility to classify the fiber-matrix connection is the push-out test. This test is basically the reversal of the pull-out test. Individual filaments are embedded in a resin layer. The individual filament is pressed out of the matrix with the aid of an indenter. As in the pull-out test, the forces measured allow conclusions to be drawn about the connection between filament and matrix. A disadvantage of this method is the usually increased variation in the results. For an initial assessment of the FMA in different matrix types the push-out test is quite suitable [4-9].

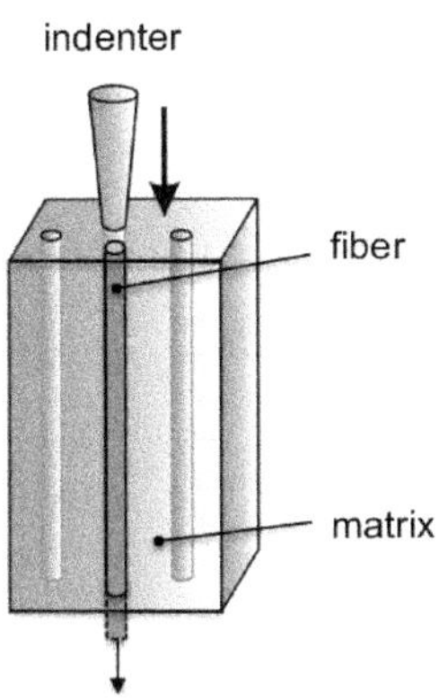

Figure 4.12
The push-out test

The advantages of both pull/push-out methods are that all types of fibers and matrix can be tested under direct measurement of the interfacial strength.

4.2.2.3 Fragmentation Test

In the fragmentation test, the single filament is embedded in a polymer matrix. The coupon sample is subjected to a tensile force until the embedded filament begins to break (fragmentation). The tensile load is maintained until no further fragmentation occurs. This marks the moment of failure of the fiber-matrix link. If there is no failure of the fiber-matrix connection, fragmentation progresses until the sample fails macroscopically. Fragmentation of the filament can be detected optically or by acoustic emission analysis. The distribution of the fragmentation lengths is recorded. The interfacial strength (FMA) can be derived under the assumption of a stress transfer model [4-10].

As with the pull-out and push-out test, the disadvantage of the fragmentation method is that only one individual filament is tested at a time and the dispersion of the individual results is correspondingly high. This means that a transfer of the results to a yarn bundle is only possible to a very limited extent.

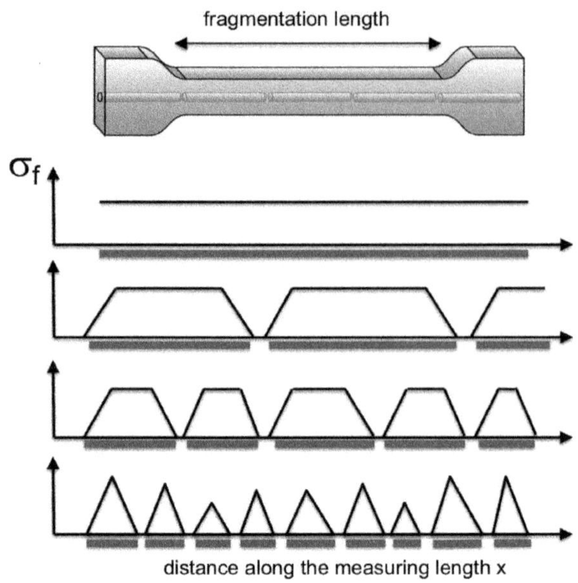

Figure 4.13 Increasing fragmentation of filaments [4-10]

4.2.2.4 Broutman Test

The Broutman test is based on the compression of a constricted molded sample in which a single filament is embedded in the longitudinal direction Figure 4.8). The application of a compressive load to the tailored specimen leads to large compressive stresses in the area of the smallest cross-section (Figure 4.14), which causes *Poisson's* expansion in the transverse direction. Since the *Poisson ratio* of the matrix is greater than that of the fiber, the transverse expansion of the matrix is greater than that of the fiber. This induces a transverse stress at the fiber-matrix interface and causes FMA failure when a critical transverse stress is exceeded. It is assumed that no friction component (friction between fiber and matrix, see pull-out test) is present when a fiber-matrix bond fails due to transverse stresses, and that the FMA can be better characterized on the basis of the results [4-11]. The separation process of fiber and matrix can be detected optically or by acoustic emission analysis.

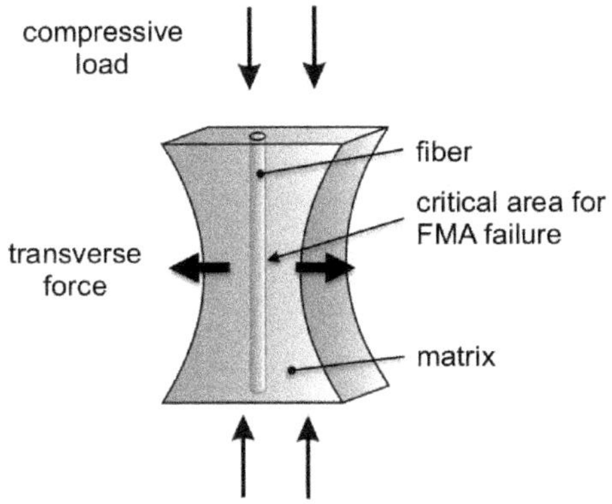

Figure 4.14
Broutman test

4.2.2.5 Testing in a Scanning Electron Microscope

The quality of a fiber-matrix connection can also be seen qualitatively using broken or already tested laminates (e.g. from a tensile or bending test) under the scanning electron microscope (SEM).

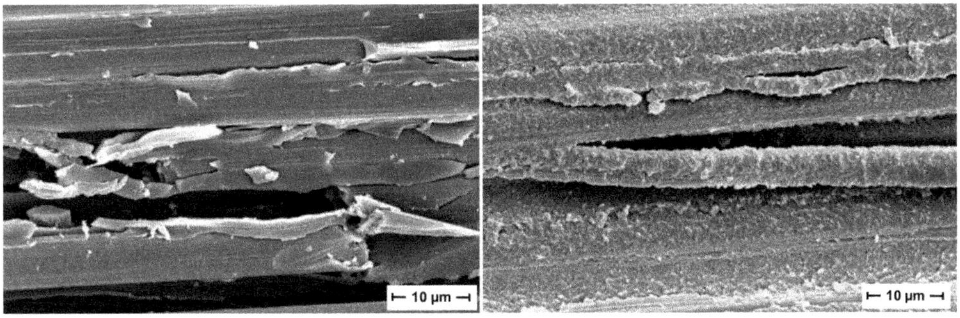

Figure 4.15 Left: poor fiber-matrix adhesion, right: good fiber-matrix adhesion
[*Arbeitskreis Neuartige Faserwerkstoffe ANF*]

In Figure 4.15, left, you can see that virtually no matrix residues adhere to the individual filaments. The filaments were peeled from the matrix as a whole, and the filament surfaces are recognizable as flat, smooth surfaces. The longitudinal texture recognizable on the filament surface corresponds to grooves, which are formed during the spinning process of the precursor and can arise through the spinnerets.

Figure 4.15, right, clearly shows matrix residues adhering to the filament surface, meaning the failure of the composite material occurred in the matrix layers. This type of failure shows the best possible combination of reactive filament surface and matching sizing and matrix.

The disadvantage of this method is that it can only provide qualitative but not quantitative information. Quantitative information can only be achieved using the aforementioned test methods.

4.3 Spectroscopic Test Methods for Carbon Fibers

Spectroscopic test methods make it possible to analyze the existing chemical bonds and the chemical structure of the carbon fiber. This is of great importance for the development of new fibers. This means that even the smallest material samples can be measured using the ratio of graphitic and diamond bonded carbon. Conclusions can be made concerning the expected fiber properties.

For chemical characterization, elemental analysis (EA), mass spectroscopy (MS), nuclear magnetic resonance spectroscopy (NMR), and Fourier transform infrared spectroscopy are mostly used.

Raman spectroscopy and X-ray photoelectron spectroscopy (XPS) are often used for structure evaluation.

In the following sections, these procedures are described in a concise manner and as briefly as possible.

4.3.1 Chemical Characterization of Carbon Fibers and Their Precursors

In order to evaluate possible precursors and to characterize carbon fibers, a number of chemical analytical methods can be used. In the following, the most important are briefly presented:

- Elemental analysis (determination of carbon weight percentage)
- Mass spectroscopy (determination of molecular fragments in a sample)
- Nuclear magnetic resonance spectroscopy – NMR (determination of molecular structure)
- Fourier transform infrared spectroscopy – IR (determination of the chemical structure)

4.3.1.1 Elemental Analysis

Elemental analysis is a technique that allows the mass distribution of chemical elements in an organic sample to be determined. Based on these data and the molar mass distribution it is possible to derive an empirical formula for the considered sample. This way the stoichiometry of an organic sample becomes accessible. With the help of further analysis methods it is then possible to derive the structural formula [4-12].

The most common form of elemental analysis is CHN analysis, which is based on the combustion of the sample. For this purpose, an exact quantity of the sample to be analyzed is weighed and then catalytically burnt at high temperatures (1800 °C - exothermic reactions) in a pure oxygen atmosphere. The combustion gases (oxidation products) are then directed to a catalytic bed of copper or tungsten with the aid of a carrier gas (in most cases helium). On this bed, nitrogen oxides (NOX) are reduced to nitrogen N_2. The reaction gases (CO_2, H_2O, SO_2, and N_2) are then separated from each other in separation columns (adsorption and desorption process) and then analyzed with the aid of a hermetic conductivity detector. The composition of the sample is then calculated using the data obtained [4-13].

In Table 4.4 typical measured values for a 100 wt.% PAN precursor fiber at selected temperatures are summarized.

Table 4.4 Typical Measured Values for PAN-Based Fibers

Final Temperature in °C:	RT	250	300
C (found)	65.88	67.22	68.30
H (found)	6.13	5.77	5.20
N (found)	25.18	25.93	24.18
C/N (calculated)	3.00		
C/N (found)	3.05	3.02	3.29
H/N (calculated)	3.00		
H/N (found)	3.38	3.09	2.99
H/C (calculated)	1.00		
H/C (found)	1.11	1.02	0.90

4.3.1.2 Mass Spectroscopy

Mass spectroscopy allows the quantitative estimation of the mass of atoms and molecules and their structure based on the fragmentation of the sample into ions, which are measured. This is possible due to the different mass/charge ratio of ionized atoms or molecules in ultra-high vacuum, which are accelerated and deflected differently in the electric field [4-14].

The general structure and functionality of a mass spectrometer are shown schematically in Figure 4.16. Each measurement is performed in three essential steps [4-15]:

- Generation of ions in the gas phase
- Separation of the ions according to their mass to charge ratio
- Quantitative measurement of ions according to mass to charge ratio

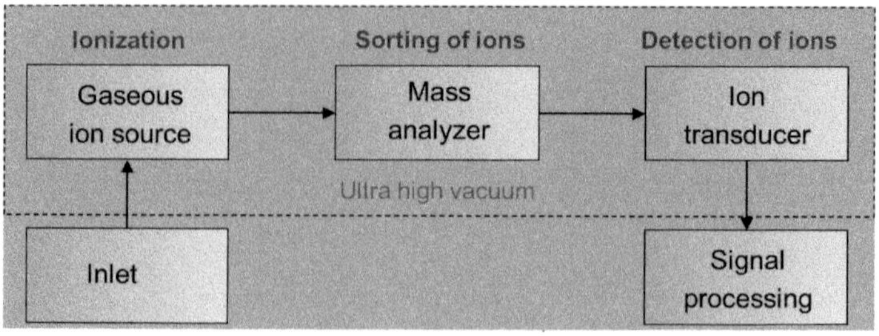

Figure 4.16 Schematic structure of a mass spectrometer

Figure 4.17 shows a mass spectrum for polyacrylonitrile as a precursor for carbon fibers.

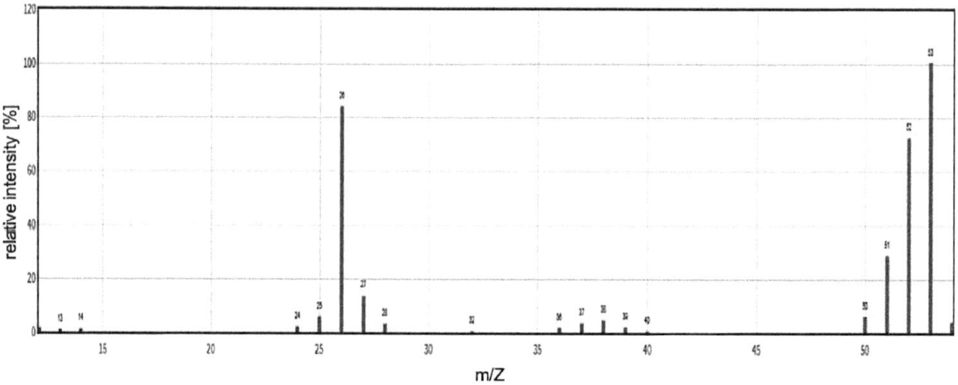

Figure 4.17 Simulated mass spectrum for polyacrylonitrile

4.3.1.3 Nuclear Magnetic Resonance Spectroscopy – NMR

Nuclear magnetic resonance spectroscopy (NMR) allows the study of minimal differences in the molecular structure of a material, making it an ideal technique for the analysis of precursors and their intermediates in the production process of carbon fibers.

The atomic nucleus of each atom has a specific rotation around its own axis, which is also called nuclear spin. The atomic nucleus of a hydrogen isotope ^{1}H, for example, is positively charged and its rotation or nuclear spin generates a magnetic moment. By applying a magnetic field it is possible to align the magnetic moments in the magnetic field. In a homogeneous magnetic field, the atomic nuclei thus have a defined energy level. To increase or decrease these energy levels, the atomic nucleus must either absorb or release energy. This phenomenon is also known as spin reversal. The energy that is absorbed or released can now be measured using nuclear magnetic resonance spectroscopy. Nuclear magnetic resonance spectroscopy thus concentrates on measuring the behavior of atomic nuclei in the magnetic field. Based on the spectra obtained, the chemical or molecular structure of the sample can then be inferred [4-16] [4-17] [4-18] [4-19] [4-20]. Figure 4.18 shows an example spectrum.

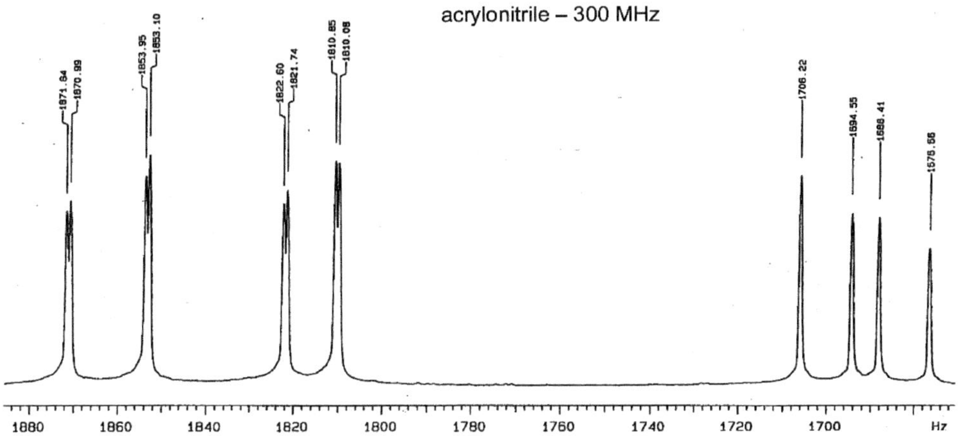

Figure 4.18 NMR spectrum of acrylonitrile - starting material for PAN based carbon fibers [4-15]

4.3.1.4 Fourier Transform Infrared Spectroscopy - IR Spectroscopy

Fourier transform infrared spectroscopy, often also more briefly called IR spectroscopy, is based on the interaction of infrared light with organic molecules. The infrared range of the electromagnetic spectrum of light has a longer wavelength and a lower frequency than the range of the spectrum visible to humans. IR spectroscopy measures the interaction of infrared light with the oscillations of atoms in molecules of a sample. The molecular structure of a sample can be derived from the spectrum [4-21].

Together with ultraviolet and Raman spectra, the infrared spectrum belongs to the molecular vibration spectra. When molecules are exposed to infrared radiation, they selectively absorb radiation of certain wavelengths, causing a change in the dipole moment of the sample molecules. Consequently, the vibrational energy levels of the sample molecules change from the ground state to the excited state. The frequency of the absorption peak is determined by the vibration energy gap. The number of absorption peaks depends on the number of vibrations of the molecule. The intensity of the absorption peaks is related to the change of the dipole moment and the possibility of the transition of energy levels. Therefore, it is possible to obtain extensive structural information by analyzing the infrared spectrum [4-21] [4-22].

Fourier transform infrared (FTIR) spectroscopy is a measurement technique used to record infrared spectra. A data processing technique known as Fourier transform converts these raw data into the desired result (the spectrum of the sample). Figure 4.19 provides an example spectrum of polyacrylonitrile, showing light transmission as a function of infrared wave number (1/wavelength).

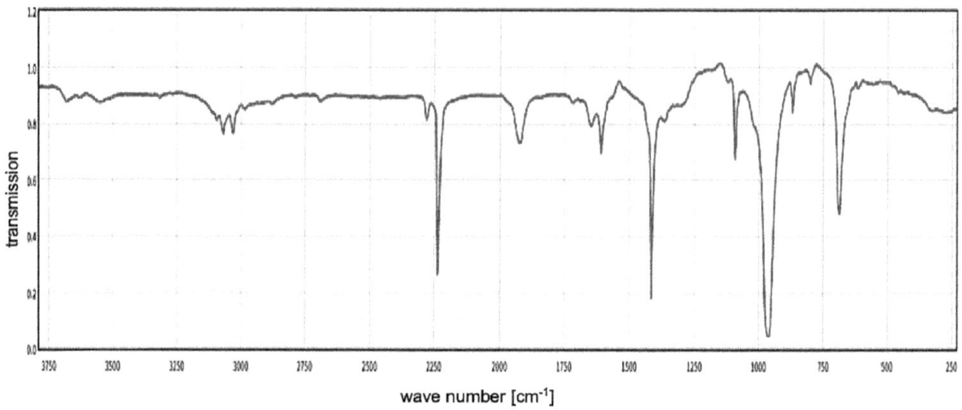

Figure 4.19 Simulated FTIR spectrum for polyacrylonitrile

4.3.2 Chemical Structure Elucidation of Carbon Fibers

After the introduction to the chemical characterization of carbon fibers and their precursors, the chemical structure will now be considered. Raman spectroscopy and X-ray photoelectron spectroscopy (XPS) are frequently used for structure elucidation.

4.3.2.1 Raman Spectroscopy

Raman spectroscopy allows the detection of different bonding forms of carbon in the carbon fiber (see Section 2.2). Thus, the quantitative distribution of sp^2- and sp^3-hybridized carbon atoms in the fiber can be detected.

The Raman effect is the physical phenomenon on which Raman spectroscopy is based. When irradiating a sample with photons from a high intensity laser, molecules begin to vibrate [4-22] [4-23] [4-24]. However, only a small part of this radiation is able to interact with the molecules under investigation, which is reflected in the Raman effect. Figure 4.20 shows the different energy levels that molecules can accept during irradiation and which types of light scattering are possible to reduce this energy [4-25].

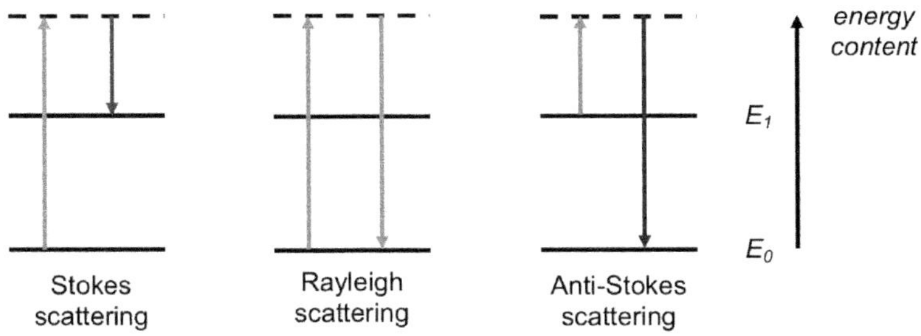

Figure 4.20 Raman effect - different energy levels

In the case of Stokes scattering and anti-Stokes scattering, the encounter between the photon and the molecule is an inelastic impact. In the case of Stokes scattering, the molecule gains energy while the photon loses energy. In the case of anti-Stokes scattering, the molecule already has a higher energy content and the energy is transferred to the photon [4-23]. *Rayleigh scattering* describes the elastic impact of a photon with a molecule, whereby the energy content remains unchanged.

The energy content in the Raman spectra is represented by the wave number [*k*] (Figure 4.22). The following equation shows how the wave number can be calculated using the speed of light [*c*] ($c = 2.99792458 \times 10^8$ m/s) and the frequency [*f*] of the photon source [4-26].

$$k = \frac{f}{100 * c}$$

For illustration, the sp^2 and sp^3 hybridization is shown in Figure 4.21. The G-band (sp^2) is described as the ring vibration of the C6 ring in carbon [4-27]. The D-band (sp^3) is characteristic for a central oscillation of a diamond structure. The direction of the sp^2 orbitals is symmetrically triangular and the sp^3 orbitals are tetrahedral.

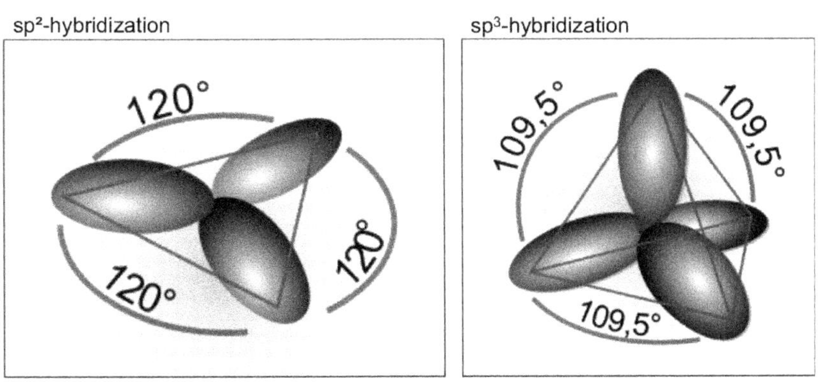

Figure 4.21 Hybridization of carbon [4-27]

The angles between the sp^2 orbitals are 120° and the angles between the sp^3 orbitals are 109.5°. Hybrid orbitals show wave functions that allow a visual idea of the spatial form of the probability of hitting an electron there [4–28]. This means that three of the four electrons in the sp^2 hybridization are bound in an atomic bond and one electron is not used for any bond. This electron is located between the graphite layers [4–29]. In the case of sp^3 hybridization, all four electrons are part of the atomic bond, which means that the carbon atom is the center of a trigonal pyramid. Figure 4.22 shows a typical spectrum for a carbon fiber.

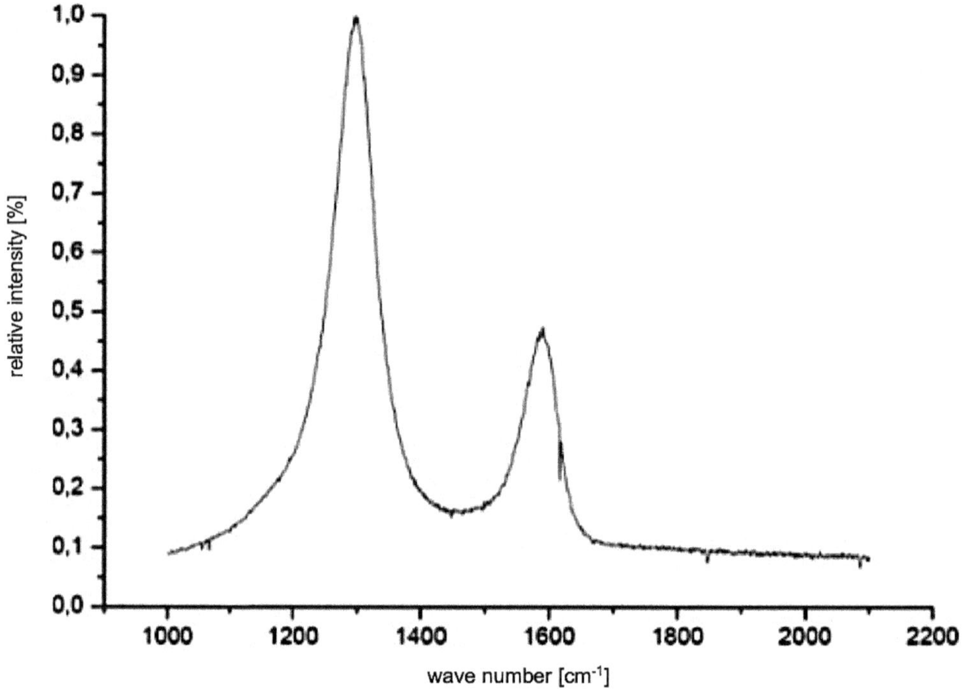

Figure 4.22 Typical spectrum for a carbon fiber

Characteristic of carbon atoms in the carbon fiber are:

- *the G-band,* which illustrates sp^2 hybridization, and
- *the D-band,* which illustrates sp^3 hybridization.

Both bands stand for the carbon modification of graphite (sp^2) and diamond (sp^3). The sp^2-hybridized G-band is typically found at a wave number of 1350 cm^{-1} and the sp^3-hybridized G-band at 1550 cm^{-1} [4–26].

A high content of sp^3 hybridization of the carbon bond (diamond structure) results in high stiffness and high tensile strength. In the diamond structure of carbon, each carbon atom is bound to four other carbon atoms. All four outer electrons of a carbon atom are involved in atomic bonds to other carbon atoms, and all atoms have the same distance from each other. In this way, four carbon atoms are bound to the central carbon atom and form a triangular pyramid, which represents the smallest part of the highly regular diamond structure. The very stable atomic bonds in this crystalline structure are the reason for the high stiffness and tensile strength of carbon fibers [4-30] [4-31] [4-32].

The anisotropic properties of a carbon fiber can be explained by the sp^2 hybridization of the carbon atoms in the carbon fiber (graphite structure). In the graphite structure, each carbon atom is bound to other carbon atoms. The resulting structure is a hexagon of six carbon atoms. These hexagons create planes. The fourth outer electron of the carbon atom in sp^2 hybridization is not involved in the atomic bonds and can move between layers, which explains the low boundary properties between these layers and the high stiffness and tensile strength within the layer [4-30] [4-31] [4-32].

4.3.2.2 X-ray Photoelectron Spectroscopy, XPS

X-ray photoelectron spectroscopy (XPS) is an important method to analyze the chemical elements contained in a carbon fiber, their chemical bonds, and their electrochemical structure.

XPS spectra are obtained by irradiating the fiber with an X-ray beam while simultaneously measuring the kinetic energy and the number of electrons escaping from the fiber to be analyzed. In this way, the conversion from precursor to carbon fiber can be illustrated by changing the elemental composition and the chemical bond [4-33].

Electrons in a chemical bond have a defined bond energy E_B, which defines the strength of a chemical bond. The binding energy is correlated with the quantum number n and the atomic number Z:

$$E_B = -13.6\ \text{eV} * \frac{Z\dagger}{n\dagger} \qquad (4.6)$$

The bond energy between two atoms is unique for the type of atoms and the type of chemical bond. The analysis of the binding energy is thus the basis of the chemical analysis with the help of XPS [4-34].

Since the energy of an X-ray beam of a certain wavelength is known and the kinetic energies of the emitted electrons can be measured, the electron binding energy of each of the emitted electrons can be determined using the photoelectric effect and the equation given in Figure 4.23. The photoelectron spectrometer detects these emitted electrons [4-35] [4-36].

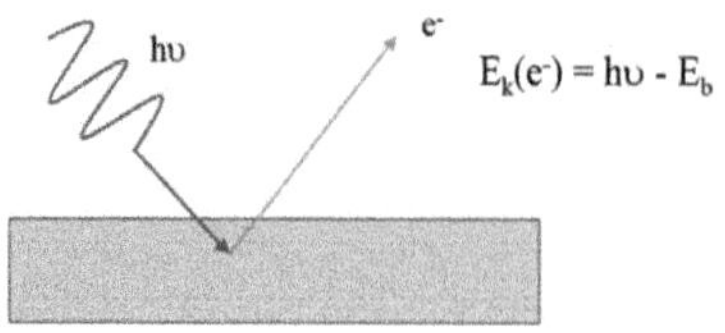

Figure 4.23
Schematic illustration of the photoelectric effect (after [4-37])

A typical XPS spectrum is an distribution of the number of detected electrons against the binding energy of the detected electrons. Each element in a carbon fiber produces a characteristic set of XPS peaks at characteristic bond energy values. The number of detected electrons in each of the characteristic peaks is directly related to the molar amount of element within the XPS sample volume [4-38] [4-39].

Figure 4.24 shows an example spectrum of a carbon fiber according to *Sherwood et al.* [4-40].

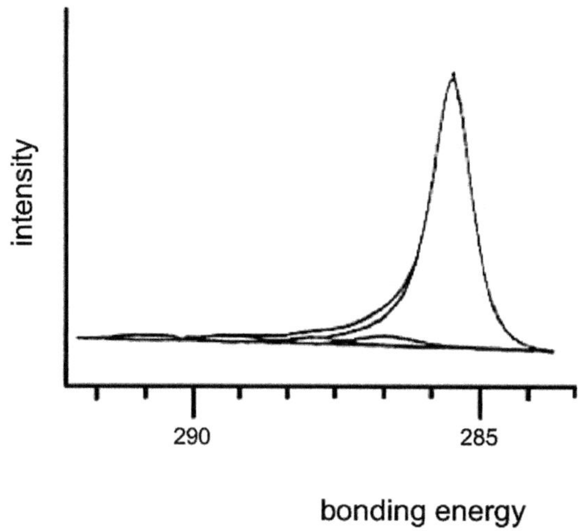

Figure 4.24
Example spectrum of a carbon fiber according to *Sherwood et al.* (after [4-40])

Outlook

In the following Chapter 5 the wide range of different applications of carbon fibers will be discussed using numerous examples.

References for Chapter 4

[4-1] W. Grellmann, S. Seidler, *Polymer Testing,* 1st edition. Munich: Carl Hanser Publishers, 2007, Chapter 10

[4-2] G. W. Ehrenstein, *Faserverbundwerkstoffe,* 2nd edition. Munich: Carl Hanser Publishers, 2006, Chapter 6

[4-3] U. Mörschel, Chr. Faymonville, E. Ingelsberger, *Automatic linear-density and tensile testing on carbon fibers,* Chemical Fibers International, 4/2008, pp. 248, 249

[4-4] FAVIMAT+, brochure and presentation of Textechno Herbert Stein GmbH & Co. KG, 2018

[4-5] ROVINGTEST, brochure and presentation of Textechno Herbert Stein GmbH & Co. KG, 2018

[4-6] E. Mäder, U. Mörschel, M. Effing, *Quality assessment of composites,* JEC Composites Magazine 102 (2016), pp. 60-62

[4-7] C. Cherif (ed.), *Textile Materials for Lightweight Constructions: Technologies - Processes - Materials - Properties,* Springer Verlag Berlin Heidelberg 2016, ISBN 978-3-662-46340-6

[4-8] FIMATEST, brochure and presentation of Textechno Herbert Stein GmbH & Co. KG, 2018

[4-9] C. Liang, J. W. Hutchinson, *Mechanics of the fiber pushout test,* Mechanics of Materials, 1993, 14, pp. 207-221

[4-10] A. Paipetis, *Composite Materials: Mechanical Behaviour & Testing,* University of Ioannina, Technological Education Institute of Serres, Greece, July 2-6, 2012

[4-11] C. Ageorges, K. Friedrich, J. Karger-Kocsis, *Fiber/Matrix Interface Testing Using The Single Fiber Broutman Test,* ICCM12 Conference Paris July 99th, Paper 234

[4-12] N. E. Mansouri, J. Salvado, *Analytical methods for determining functional groups in various technical lignins.* Industrial Crops and Products, 2007, 26, pp. 116-124

[4-13] J. Liebig: *Über einen neuen Apparat zur Analyse organischer Korper, und die Zusammensetzung einiger organischer Substanzen [About a new apparatus for the analysis of organic bodies, and the composition of some organic substances].* Ann. Phys., 1831, 21, pp. 1-47

[4-14] A. G. Marshall, C. L. Hendrickson, *High-resolution mass spectrometers.* Annu. Rev. Anal. Chem. 2008, 19, pp. 1-21

[4-15] L. Chen, L. Bromberg, H. Schreuder-Gibson, J. Walker, T. Hatton, G. Rutledge, *Chemical protection fabrics via surface oxidation of electrospun polyacrylonitrile fiber mats.* J. Mater. Chem., 2009, 19, pp. 2432-2438

[4-16] Aue, W., E. Bartholdi, R. R. Ernst, *Two-dimensional spectroscopy. Application to nuclear magnetic resonance.* The Journal of Chemical Physics, 1976, 64, p. 2229

[4-17] Gomathi, L., *Elucidation of secondary structures of peptides using high resolution NMR.* Current Science, 1996, 71(7), p. 553

[4-18] Ames, J. B., Hamasaki, N., Molchanova, T., *Structure and calcium-binding studies of a recoverin mutant (E85Q) in an allosteric intermediate state.* Biochemistry, 2002, 41(18), p. 5776. DOI: 10.1021/bi012153k

[4-19] R. R. Ernst, W. A. Anderson, *Application of Fourier transform spectroscopy to magnetic resonance.* Review of Scientific Instruments, 1966, 37, pp. 93-201

[4-20] W. P. Aue, E. Bartholdi, R. R. Ernst, *Two-dimensional spectroscopy. Application to nuclear magnetic resonance.* The Journal of Chemical Physics, 1976, 64, pp. 2229-2246

[4-21] K. P. C. Vollhardt, N. Schore, *Organic Chemistry: Structure and Function,* New York: W. H. Freeman, 2009, pp. 468-503

[4-22] V. Brückner, *Elements of Optical Networks: Basics and Practice of Optical Data Transmission,* Vieweg + Teubner Verlag, 2011

[4-23] W. Demtröder, *Laser Spectroscopy,* Vol. 1, Springer Verlag, 2011

[4-24] J. Eichler, *Physik fur das Ingenieurstudium,* Vieweg + Teubner Verlag, 2011

[4-25] Z. B. Hesse, *Spectroscopic Methods in Organic Chemistry,* Stuttgart: Georg Thieme Verlag, 2008

[4-26] S. B. Hornbogen, *Mikro- und Nanoskopie der Werkstoffe,* Springer Verlag, 2009

[4-27] H.-J. Bargel, *Werkstoffkunde,* Berlin: Springer Verlag, 2008

[4-28] H. Emminger, *Physikum,* Stuttgart: Thieme Verlag, 2003

[4-29] H. P. Latscha, *Anorganische Chemie, Chemie - Basiswissen,* Berlin: Springer, 1992

[4-30] J. G. Graselli, *Analytical Raman spectroscopy,* John Wiley & Sons Inc., 1991

[4-31] T.J. Dines, D. Tither, A. Dehbi, A. Matthews, *Raman spectra of hard carbon films and hard carbon films containing secondary elements.* Carbon, 1991, 29(2), pp. 225-231

[4-32] L. Tuinstra, *Raman spectrum of graphite,* The Journal of Chemical Physics, 1970, 53, p. 3

[4-33] J.J. Yeh, I. Lindau, *Atomic subshell photoionization cross sections and asymmetry parameters.* Atomic Data and Nuclear Data Tables, 1985, 32, pp. 1-155

[4-34] D. Briggs, M.P. Seah, *Practical Surface Analysis,* Vol. 1, Wiley, 1990, MaWi-Bibliothek (Nro.668)

[4-35] G. Ertl, J. Küppers, *Low Energy Electrons and Surface Chemistry,* Wiley-VCH, 1985, (No. 962)

[4-36] J.C. Riviere, *Surface Analytical Techniques,* Clarendon Press Oxford, 1990, MaWi-Bibliothek (No. 494)

[4-37] W. Göpel, C. Ziegler, *Struktur der Materie,* Teubner, Leibniz 1994

[4-38] J. Chastain, R.C. King (ed.), *Handbook of X-Ray Photoelectron Spectroscopy,* Physical Electronics

[4-39] M. Grasserbauer, H.J. Dudek, M.F. Ebel, *Angewandte Oberflachenanalyse,* Springer 1985, MaWi Library (No. 299)

[4-40] P.M.A. Sherwood, *Surface analysis of carbon and carbon fibers for composites,* Journal of Electron Spectroscopy, 1996, 81, pp. 319-342

5 Applications of Carbon Fibers

5.1 Introductory Remarks

Due to their excellent and versatile properties, PAN-based fibers are used in a wide variety of applications. Application examples of different fiber types (PREOX and carbon fibers) and different fiber format (continuous, cut, and ground) will be explained in this chapter.

The diagram in Figure 5.1 shows a variety of dry and pre-impregnated fiber semi-finished products, starting with PAN fiber as precursor followed by PREOX fiber and carbon fiber. Furthermore, different possible processing technologies producing different end-products are presented.

Starting from a continuous PAN precursor fiber whose *K number* can be varied depending on the end application (see also Chapter 3), a so-called PREOX fiber (oxidatively stabilized PAN fiber) is produced as the first step in the carbon fiber production process. This PREOX fiber can either be directly converted into carbon fiber, or can be used in textile processing steps in the form of continuous or short cut fiber or fiber flour. The respective K number of the precursor depends mainly on the application of the fiber. It influences the property level and thus also the fiber price considerably. For example, while only low K numbers of 3 to 24 K are used for aerospace applications, yarn counts of 50 to 400 K are common for industrial and automotive applications.

Depending on the application, the fibers are further processed either as dry textiles (with subsequent matrix addition) or as semi-finished products pre-impregnated with matrix (e.g. thermoplastics). The diagram in Figure 5.1 shows how the wide range of semi-finished products can be processed into a finished component in a wide variety of ways.

The following examples show applications of carbon fibers and their processing into components and end-products.

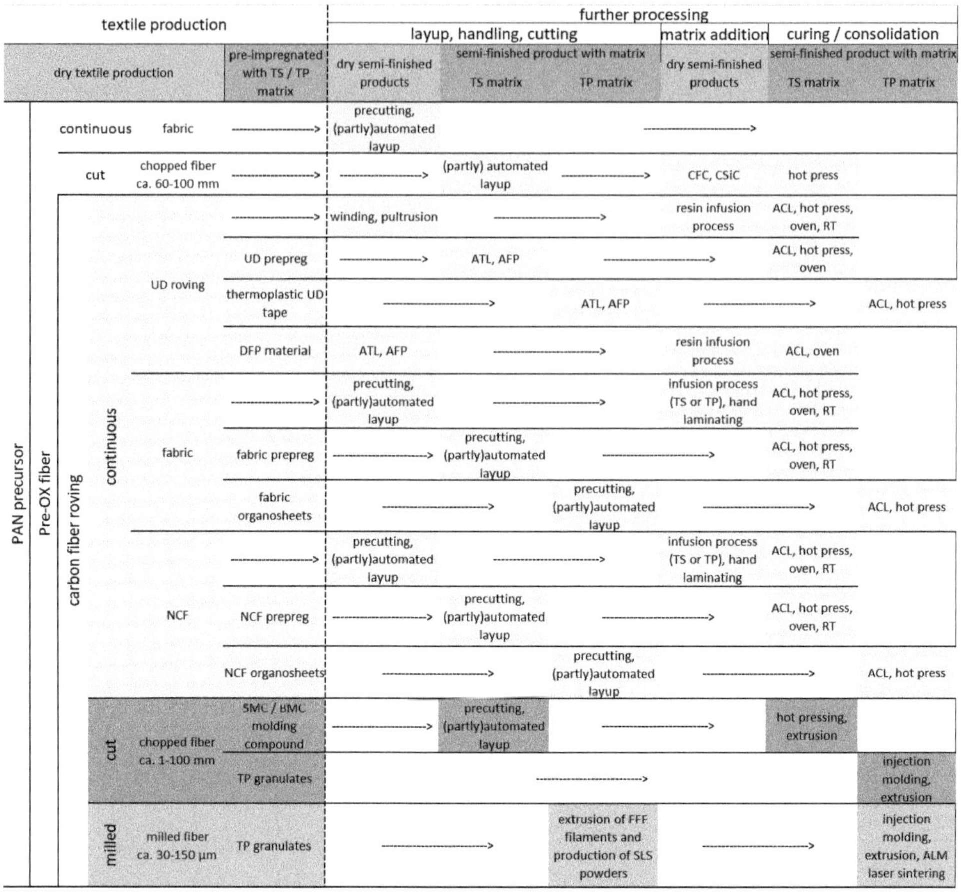

Figure 5.1 From precursor to application

■ 5.2 Dry Semi-Finished Fiber Products

5.2.1 PREOX Fibers

As described in Chapter 3, PREOX fibers or yarns are a thermally stabilized intermediate product in the carbon fiber production process. These can either be further processed into carbon fibers or used directly as continuous fibers or as crimped staple fibers. PREOX fiber types commonly offered on the market are summarized in Table 5.1.

Table 5.1 PREOX Fibers

Manufacturer	Type Designation	Presentation
SGL Composites	PANOX®	continuous fiber, crimped staple fiber
Toho Tenax Europe	PYROMEX®	continuous filament, crimped staple filament
ZOLTEK Corporation	ZOLTEK OX™	continuous fibers, stretch-broken yarns, staple fibers, fabrics, fleeces

These stabilized PAN fibers have numerous interesting properties and a large number of different products have been developed. In addition to continuous fibers, which can be further processed with classic textile processes after appropriate surface pretreatment (sizing), there are other typical product forms:

- weave
- nonwovens
- staple fibers
- short fibers
- spun yarns from broken fibers (so-called stretch broken yarns)
- fiber flours

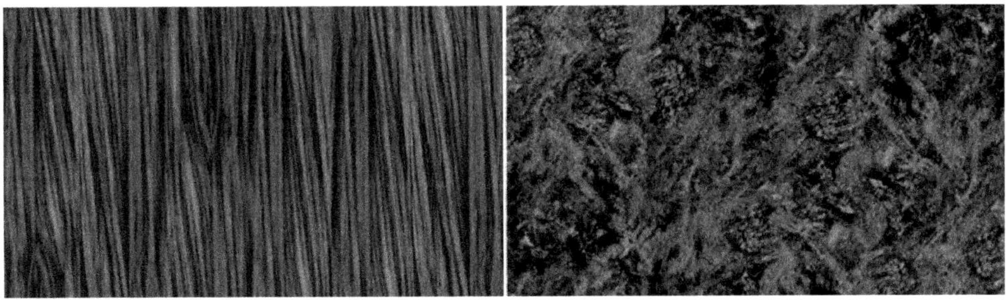

Figure 5.2 Left: PYROMEX® continuous fiber, right: PYROMEX® staple fiber [Tejin Carbon]

Crimped staple fiber is usually cut to lengths between 50 and 100 mm. The typical K number for staple fibers is around 320,000 filaments (320 K) in the fiber strand.

The applications for PREOX yarns are as varied as the dosage forms; PREOX fibers are used, for example, in the following:

- Refractory building and aircraft interiors, e.g. as fabrics in seats for aircraft and public buildings
- Fire and heat protective clothing, also in combination with other fibers, e.g. for gloves and protective suits for the fire brigade, the steel industry, and motor racing

- Fire-resistant fabric or fleece inserts in welded ceilings, fire protection and acoustic barriers in the automotive sector
- Ground PREOX fiber as a component in asbestos-free friction linings
- Starting material (short fibers) e.g. for aircraft brake discs made of carbon fiber-reinforced carbon (CFC) or in the automotive industry for carbon ceramic brake discs (CSiC)

CFC as a secondary product is characterized in particular by its thermal shock stability and high stiffness while maintaining its strength up to a temperature of around 1600 °C. The disadvantage is its oxidizability, which can, however, be reduced by suitable additives. CFC can be converted into a ceramic composite material by siliconization in suitable processes to produce carbon fiber-reinforced silicon carbide (CSiC).

Typical applications for CFC are aircraft brake discs (see Figure 5.3) and rocket nozzles, for which thermal shock stability above 1500 °C is required. The extreme temperature and oxidation stability of up to 1100 °C and very low thermal expansion make CSiC an ideal material for thermally highly stressed structures in aerospace applications and for high-performance brake discs in the automotive sector. Compared to conventional cast iron brake discs, CSiC brake discs are significantly lighter (around 50%).

Figure 5.3
Aircraft brake made of PANOX® oxidized fibers [*SGL Technologies GmbH*]

Figure 5.4 Audi carbon-ceramic brake system [*Audi-Mediacenter*]

5.2.2 Carbon Fibers

5.2.2.1 Carbon Fibers, Continuous

Continuous carbon fibers offer material properties such as tensile strength and stiffness at the highest level and are therefore ideally suited for mechanically highly stressed components. They form the basis for load-bearing primary structures in aircraft construction, transportation, and the energy sector, as well as in many other areas of application where an extraordinary mechanical property level is required with low weight at the same time.

UD Rovings

Unidirectional carbon fibers as roving or tow can be further processed in very different ways as shown in the diagram of Figure 5.1.

Winding and pultrusion processes are ideal for processing dry carbon fibers directly into components. Both processes offer the possibility of manufacturing components in large quantities and with very good precision. Especially the winding process allows the possibility to process fibers in dry as well as in pre-impregnated or in-situ impregnated condition using a forming tool. Typical components produced by the winding process are pressure tanks, for example for solid-fuel rockets for space travel (Figure 5.5) or for novel hydrogen tanks for fuel cell-powered vehicles (Figure 5.6). The four hydrogen tanks of the Audi A7 Sportback h-tron quattro are located in front of the rear axle and in the center tunnel of the vehicle. An outer skin of carbon fiber reinforced plastic (CFRP) surrounds the inner aluminum shell. Up to 5 kilograms of hydrogen can be stored in the tanks at a pressure of 700 bar - enough to travel over 500 kilometers.

Figure 5.5 Demonstrator of a tank for solid-fuel rockets for the aerospace industry, manufactured in the winding process [*MT Aerospace AG*]

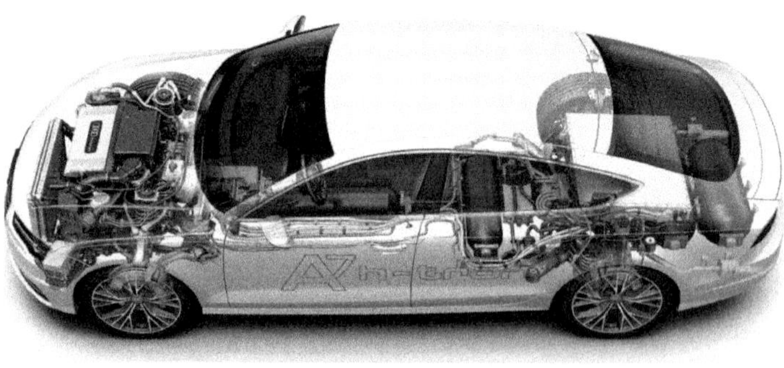

Figure 5.6 Audi A7 Sportback h-tron quattro [*Audi-Mediacenter*]

Another automotive application of UD tapes is shown in Figure 5.7. The Audi A8 rear panel is made from 50 K carbon fibers. The unidirectional tapes are stretched, cut, and arranged into a tailor-made preform. This preform is transferred into the Ultra-RTM process by *Voith Composites* and further processed to the final back wall.

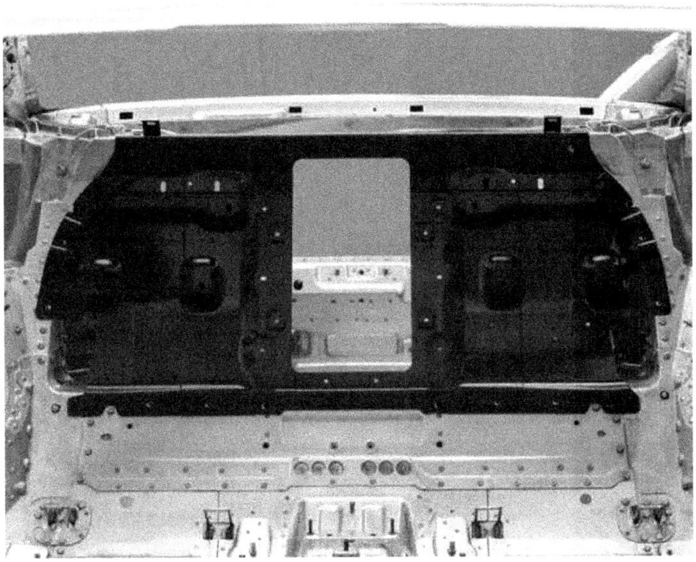

Figure 5.7 Audi A8 rear panel made of 50 K carbon fibers
[*Audi Media Center, Voith Composites*]

Figure 5.8 shows a thick-walled CFRP flywheel developed by *Schunk Kohlenstoff-technik* and manufactured using wet winding technology. The CFRP flywheel can be used to store kinetic energy in a KERS (Kinetic Energy Recovery System) in trucks and commercial vehicles. This lightweight product with very high stiffness in the circumferential direction allows storage of much more energy than a similar metallic component. The reason for this is that the stored energy grows with the square of the speed of rotation, but only linearly with the mass. Metallic flywheels do not withstand the high centrifugal forces at high rotational speeds due to the four times higher density compared to the CFRP component. The technology can be installed in many vehicles and can also be used to store electrical energy in purely electric vehicles, making it a technology of the future.

While such flywheels have been used in motor sports for quite some time, the commercial product is currently still in the test phase and should be launched on the market in 2021.

Figure 5.8 Wrapped CFRP flywheel [*Schunk Kohlenstofftechnik GmbH*]

Whereas during winding the outer component shape must be rotationally symmetrical (round, square rounded), the pultrusion process offers extended design freedom. Undercuts or sharp-edged components can be produced in large quantities. In addition, both processes can also be combined with each other in the pull-winding process. All profiles are characterized by low weight and high strength compared to metallic variants.

Figure 5.9 UD fibers and epoxy resin matrix – left: hollow chamber profile, right: profile with undercuts [*Secar Technologie GmbH*]

A future application in automotive engineering could be the use of carbon fiber pultrusion profiles as side impact beams. The pultrusion process or drawing mesh is used for profile parts such as the side impact beam, e.g. from the *Brose* company. Here it is possible to produce straight and curved profiles in a continuous process at comparatively low tool costs. Since both methods produce practically no waste, material costs can also be reduced. Figure 5.10 shows a concept for a *Brose* side impact beam published at the IAA in 2013.

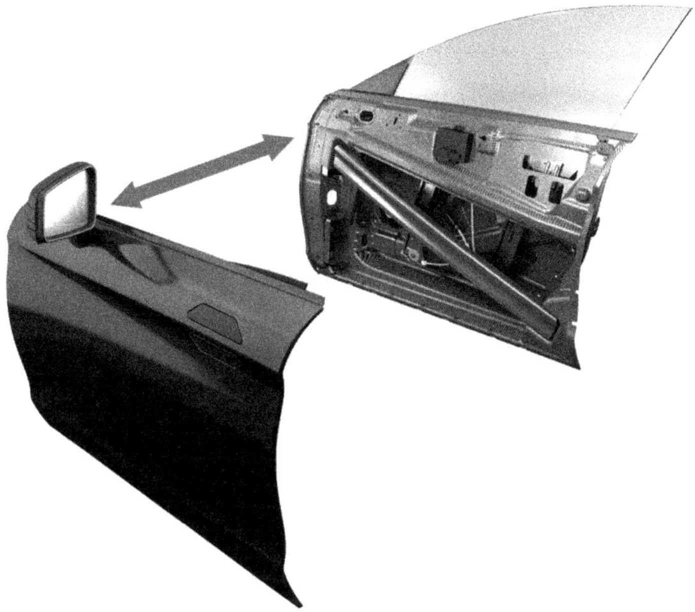

Figure 5.10 Carbon fiber pultrusion profile as side impact beam (IAA press release Frankfurt/Main, September 11, 2013: [*Brose*], CFRP door structure)

In order to make round UD pultrusion profiles also suitable for torsion loads, they can be wrapped with continuous carbon yarn in the pull-winding process in the circumferential direction at various angles (Figure 5.11).

Figure 5.11 Pull-winding profiles made of UD fibers and epoxy resin [*Secar Technology*]

Figure 5.12 left shows a CFRP lamella, which is also produced by pultrusion from UD carbon fibers and corrosion-resistant vinyl ester resin. These lamellas are used in the construction industry for strengthening or repairing structures (cracks), e.g. bridges or multi-story buildings (Figure 5.12, right). The lamellas are glued to the pre-treated concrete surface from the outside using a special adhesive. Mechanical anchoring to the structure by means of clamps is also possible in principle. The slats can be applied in a flaccid or pre-stressed state and then absorb the resulting stresses in the structure and prevent further crack growth in reinforced concrete structures.

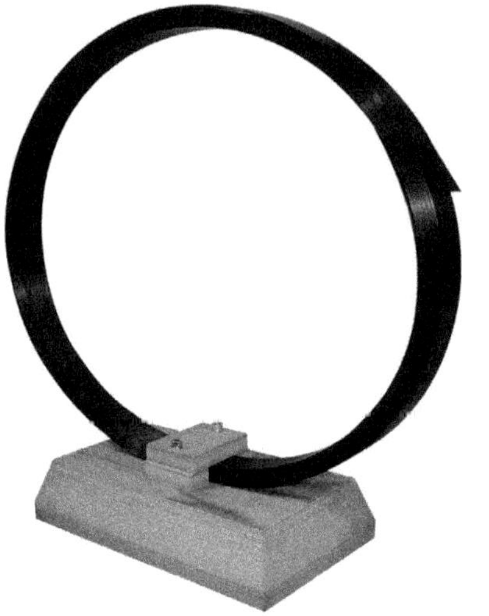

Figure 5.12 CFRP lamella for construction industry made of UD fiber and epoxy resin matrix; left [*Röchling Engineering Plastics SE & Co. KG*], right [*Prof. Martin Betzler, hochschule21*]

With this technology, buildings can be upgraded, for example if a building is to be subsequently extended by an additional story or if a change in use is planned and the existing statics are no longer sufficient. In the case of damaged buildings, a repair can be carried out instead of expensive renovation or even demolition/new construction. CFRP slats are an excellent alternative to steel slats as they are considerably lighter than steel slats. They are easier to handle and can be glued by construction personnel on site.

The company *Schunk Kohlenstofftechnik* manufactures two very interesting products from continuous carbon fibers, in which the high temperature strength and the electrical conductivity play a central role and at the same time the high stiffness and the potential of the carbon fiber are exploited. Figure 5.13, left, shows a transport chain for high-temperature applications, which has been on the market for more than 15 years and is used in the glass industry or in belt furnace processes under inert atmospheres. This is used for the transport of products in high-temperature processes up to 2000 °C under protective gas or in vacuum, where high tensile strengths at application temperatures of > 1000 °C, low thermal expansion, heat capacity, and freedom from contamination are required. The chain links and axles are manufactured using wet winding technology with phenolic resin and then carbonized at 1100 °C, then graphitized at 2000 °C. The chain links and axles are then coated with a graphite coating. In the final assembly, the two components are joined to form the complete CFC chain using ground graphite rollers and round CFC plates. In addition to the advantages mentioned above, the chain shows very little distortion at high temperatures (unlike metallic materials), excellent wear resistance, and is chemically inert. The CFC material can be additionally siliconized or coated with pyrocarbon (PYC), silicon carbide (SiC), silicon (Si), or silicon nitride (Si_3N_4) in the CVD process to give the product even better chemical stability.

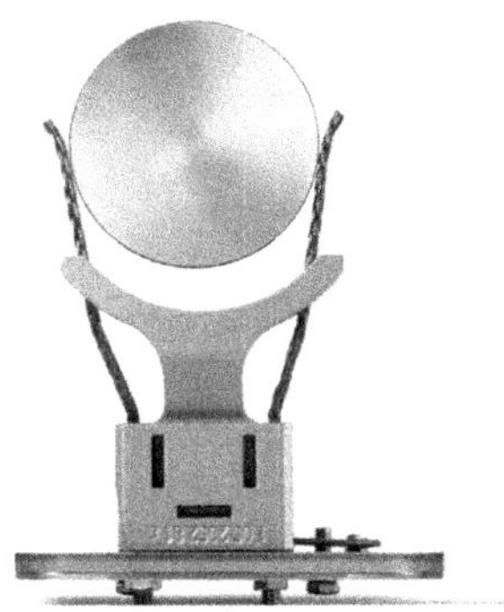

Figure 5.13 Left: CFC transport chain for high-temperature applications, right: carbon fiber wave earthing [*Schunk Kohlenstofftechnik GmbH*]

Figure 5.13, right, shows a shaft grounding system from *Schunk Kohlenstofftechnik*, a so-called carbon-fiber grounding system (CFG system) for the derivation of high-frequency currents (kHz-MHz) for electric motors in railway and industrial applications. These systems protect ball bearings against wear caused by spark erosion and thus ensure fail-safe mobility in many rail vehicles. Each fiber strand of a contact grinder consists of up to 200,000 individual fibers which reliably dissipate the high-frequency current. To manufacture the systems, carbon fibers are braided and infiltrated with pyrolytic carbon, and then joined with a connecting element. Infiltration with pyrocarbon increases the wear resistance and ensures optimum electrical connection between the individual fibers. The fiber bundle obtained is so wear-resistant that it can be used for a distance of up to 500,000 km before the system has to be replaced. This technology of shaft grounding can be regarded as a hidden champion and real problem solver in the mobility industry, as it saves more than € 100,000 (ca. US$ 115,000) per vehicle in maintenance costs every year, which would otherwise be incurred for the replacement of ball bearings and the associated maintenance effort.

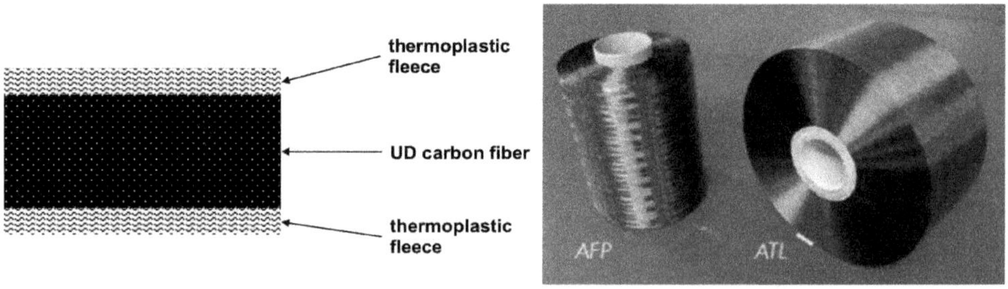

Figure 5.14 Left: structure of HiTape, right: configuration as AFP and ATL material [*Hexcel Corporation*]

Dry fiber placement (DFP) tapes have been on the composite market since 2014 (Figure 5.14). They are manufactured on the basis of dry carbon fibers and a thermoplastic fleece coating. The thermoplastic fleece used for the coating usually has a weight per unit area of only 4 to 10 g/m^2 and is located exclusively on the surface (top and bottom) of the UD roving. DFP tapes can thus be regarded as both dry and pre-impregnated semi-finished products. The thermoplastic fleece has three essential functions:

1. It serves as a thermally activated binder between the DFP tape and the mold surface or between the individual tape layers.
2. In the subsequent impregnation process, the nonwoven ensures better and faster impregnability with a reactive resin (e.g. RTM6) as spacer between the layers.
3. In the cured component, the thermoplastic nonwoven acts as an insoluble modifier between the layers to increase the system's crack toughness (CAI parameters) and functions in the same way as the insoluble particles in third-generation prepregs.

The level of mechanical properties of a DFP component with RTM6 as the matrix system is now comparable to that of an aerospace prepreg system such as HexPly® M21 due to the introduction of thermoplastic nonwovens.

Depending on the tape width, DFP tapes are processed either by fiber placement (AFP, e.g. width 6.3 mm) or automatic tape laying (ATL, e.g. width 300 mm). Furthermore, the material can be processed just like the dry carbon fibers mentioned above using the winding and pultrusion technology. The narrow material can even be used in UD braiding technology. Impregnation with a reactive thermoset system always takes place after the DFP tape has been placed in the mold. This eliminates the cost-intensive cooling of the material required for many prepregs.

Table 5.2 DFP Tapes - Manufacturers and Systems

Manufacturer	Designation DFP Tape	Infusion Resin System
Hexcel Composites	HiTape®	HexFlow® RTM6 / RTM6-2
CYTEC Solvay Group	PRISM® TX1100	PRISM® EP2400

The DFP technology is still young, so there are currently no mass-produced components. Components to be expected in aviation will probably be wing and fuselage components for new single-aisle aircraft. An important reason for the change from prepreg to DFP technology is the possibility to reduce the total production costs for composite components based on UD tapes. Thus, the time- and energy-intensive process of autoclave curing can be dispensed with and instead curing can take place in a hot-air oven exclusively under vacuum. The process flow is shown in Figure 5.15.

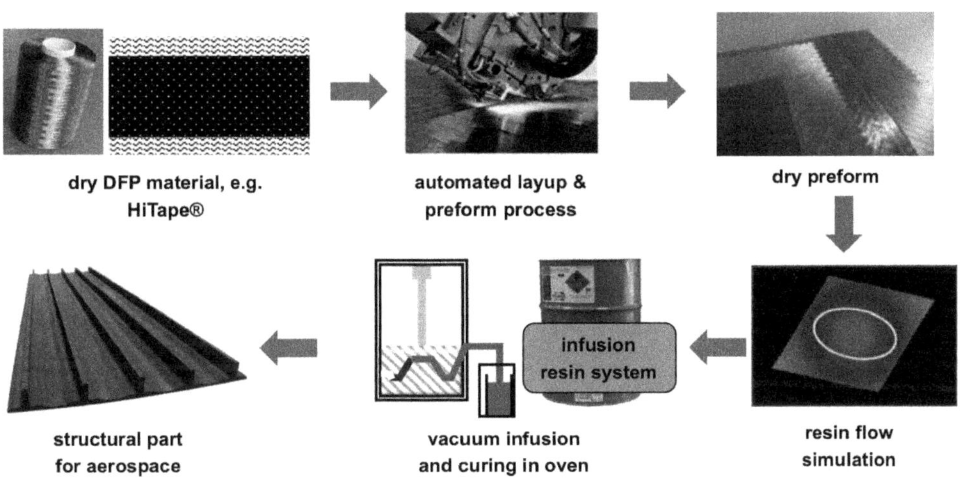

Figure 5.15 Process flow from DFP tape to component [*Hexcel Corporation, DLR Brunswick*]

Figure 5.16 Omegastringer-strengthened panel based on DFP tape and RTM resin [*Hexcel Corporation*]

Fabric Weaves and Non Crimped Fabrics (Multiaxials)

Continuous carbon fibers – like other textile fibers – can be processed excellently after suitable surface treatment (sizing, see Chapter 3) with the aid of textile processes into fabrics and multiaxial, non-crimped fabrics (NCF). Fabric weaves are probably the oldest textile structures for industrial fibers, such as carbon, glass, or aramid fibers. By using different types of weave, the mechanical properties can be perfectly adapted to the respective component application in the subsequent compound with the matrix resin. Even different fibers can be mixed together in one fabric or NCF. Figure 5.17 shows three different types of weaving. It can be seen very clearly that the weft thread is undulating (deflection of the fiber). This is called undulation. The crimp factor depends on the one hand on the type of weave (canvas weave = strong undulation, atlas weave = low undulation) and on the other hand on the *K number* of the roving used. Both factors influence the process ability of the dry semi-finished product as well as the mechanical property level in the subsequent composite.

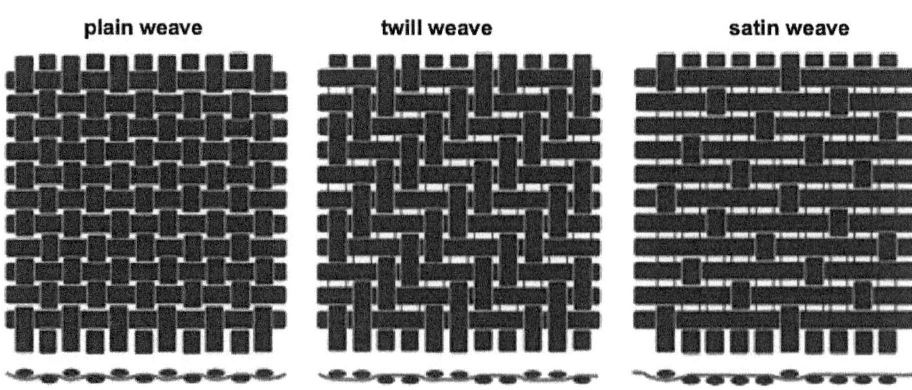

Figure 5.17 Examples of weave types of fabrics

In order to reduce the undulations (crimp factor) in fabrics to improve the mechanical properties, UD rovings are spread before or during fabric manufacture, i. e. the individual yarn (tow) becomes wider and thinner at a given *K number*, so that the crimp decreases at the crossing points. Another advantage of this technology is that very light fabrics (< 100 g/m^2) can be produced relatively easily. In addition, the open areas at the crossing points in the fabric are almost completely closed, resulting in an almost completely closed fabric appearance (Figure 5.18). This is ideal for visible carbon components as well as for honeycomb components. Spreading also represents the possibility of reducing the cost of fabrics, as this technology allows the use of UD rovings with a higher K number. at lower raw material costs. This means that UD rovings with a higher K number can be used for a constant basis weight of the fabric. Depending on the procedure, increased processing costs may have to be taken into account. Suppliers of spread tow fabrics include *Hexcel* (PrimeTex®) and *Oxeon* (Textreme®).

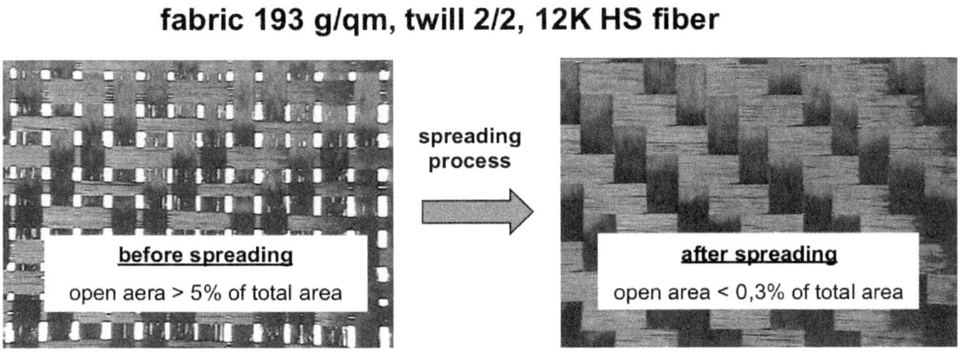

Figure 5.18 Spread tow fabric PrimeTex® [*Hexcel Corporation*]

Figure 5.19 Left: plain weave standard, right: Textreme® spread tow fabric

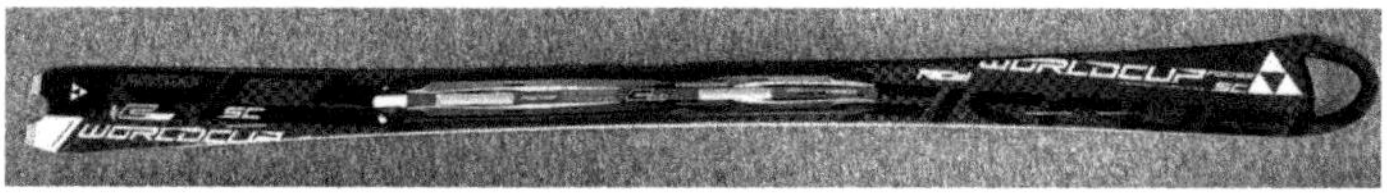

Figure 5.20 Application of PrimeTex fabric for winter sports skis [*Fischer Ski*]

While the fibers in fabrics undulate to different degrees depending on the type of weave (Figure 5.17), the fiber bundles in multiaxial fabrics (NCF) are available as stretched yarn layers (Figure 5.21). This enables optimum mechanical strength, very high drape-ability and excellent permeability in the subsequent infusion process. At the same time, the use of multiaxial fabrics allows cost savings to be achieved by reducing the laying effort due to the lower number of layers in the component with a high single layer basis weight during component manufacture (Figure 5.23).

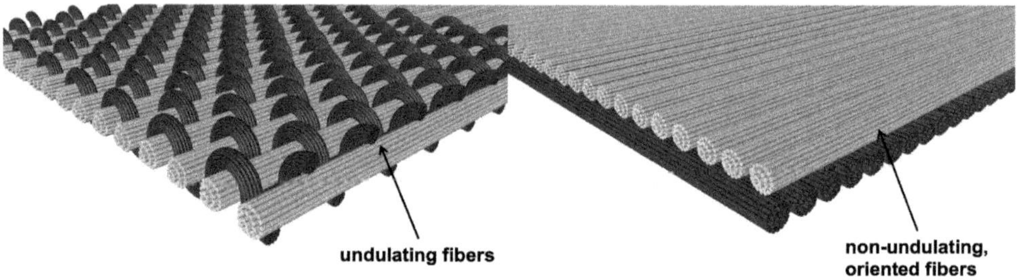

Figure 5.21 Differences in thread position in fabric (left) and multiaxial fabric (right) [*SAERTEX GmbH & Co. KG*]

Based on this technology, *SAERTEX* and *Audi* have set themselves the task of developing a carbon solution for the B-pillar of the Lamborghini Huracán and Audi R8. Carbon NCFs from SAERTEX were used for this application. The advantages of carbon NCFs are high drape-ability, good resin permeability, and excellent quality assurance in production. Figure 5.22 shows the Audi space frame in multi-material construction with carbon solution for the B-pillar of the Lamborghini Huracán and Audi R8.

In Figure 5.23 the number of individual layers, the layer orientation, as well as the fiber weight per unit area (K number of the individual roving) can be individually adjusted in the scrim. The applications for scrims are therefore very diverse due to this high adaptability to the respective requirements and the special properties. The following illustrations show various applications for woven fabrics and multiaxial fabrics.

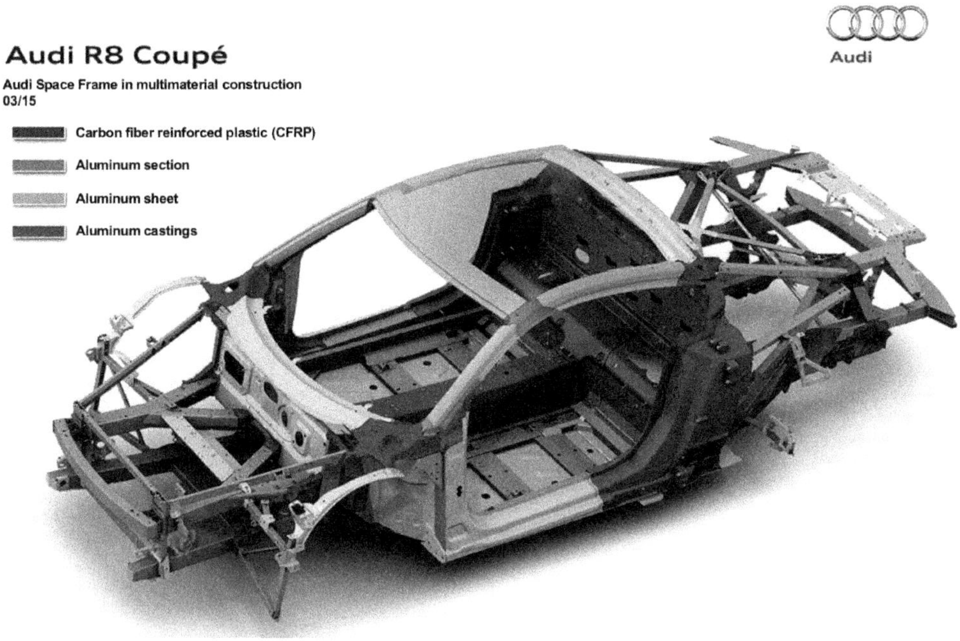

Figure 5.22 Audi space frame in multi-material construction with carbon solution for the B-pillar of the Lamborghini Huracán and Audi R8 [*Audi Media Centre*]

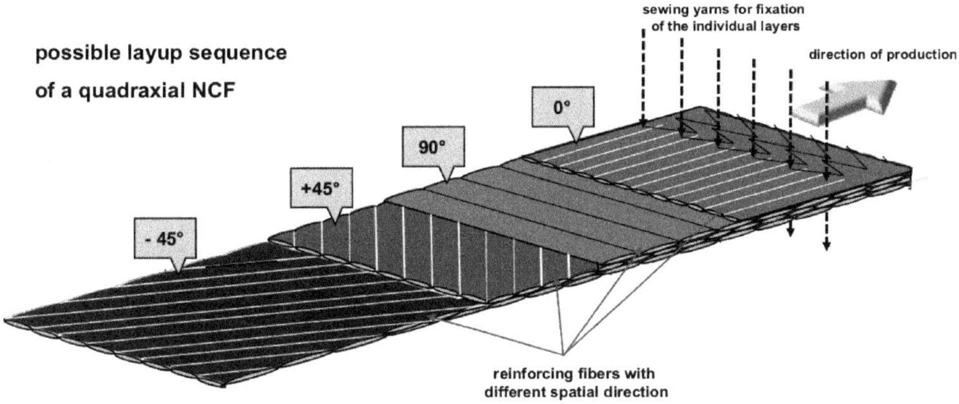

Figure 5.23 Structure of fabrics - combination of up to four individual layers

Figure 5.24 shows the result of a concept study for the development of a monolithic design for a spoiler of a commercial aircraft. In addition to manufacturing aspects such as liquid composite molding (LCM) technology and out-of-autoclave curing, special attention was paid to the rear stiffening structure during design. Using topology optimization, a structure modeled on nature (principle of a bird's wing) was developed in which the force introduction area required for connecting the hydraulic actuator (fitting, Figure 5.24, right) is integrated. The fabric used here offers optimum drape-ability for implementing this design in the component and for producing a preform, which is then impregnated with epoxy resin using LRI technology (liquid resin infusion) and cured.

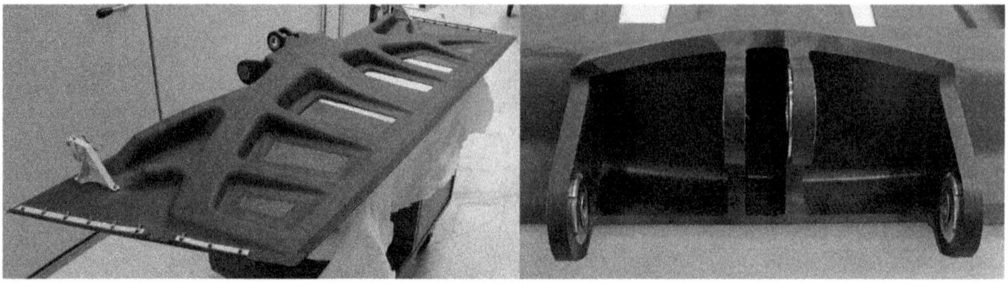

Figure 5.24 Fabric application in concept study for the monolithic construction of an aircraft spoiler [*FACC AG*]

Figure 5.25 shows a biaxial multiaxial fabric as used for the production of double-curved structures such as a pressure dome in aviation.

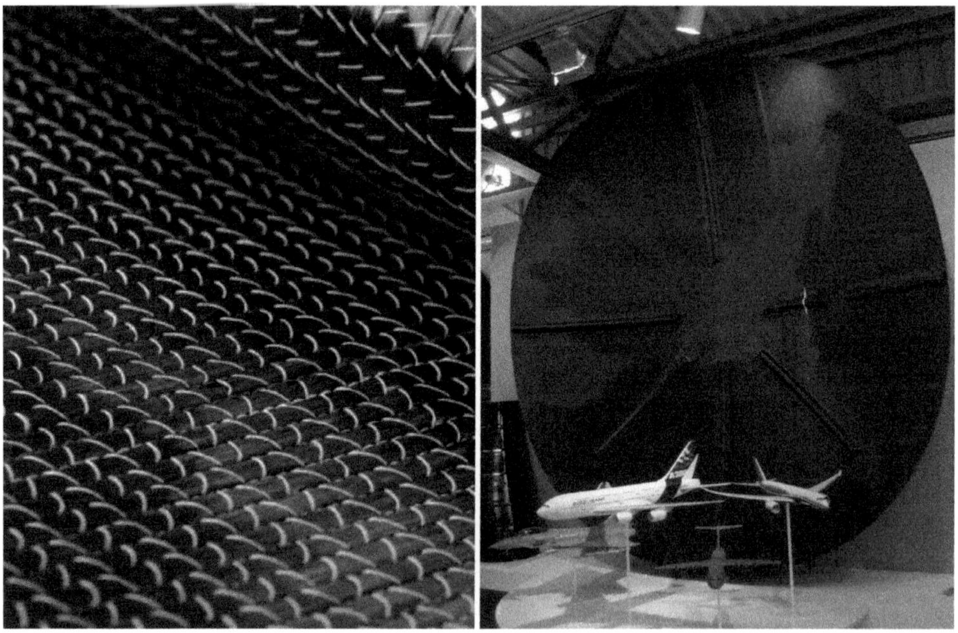

Figure 5.25 Application of multiaxial fabrics in aviation – demonstrator of a rear pressure bulkhead [*Hexcel Corporation*]

In the construction industry, wide-meshed mesh fabrics made of UD carbon fibers are increasingly being used as reinforcement for concrete in place of classic reinforcement steel meshes. Carbon fiber mats offer advantages over the metallic variant such as corrosion resistance, weight, and higher load-bearing capacity under tensile and compressive loads. In addition, the mats are considerably thinner, so that the overall thickness and weight of a concrete floor decrease.

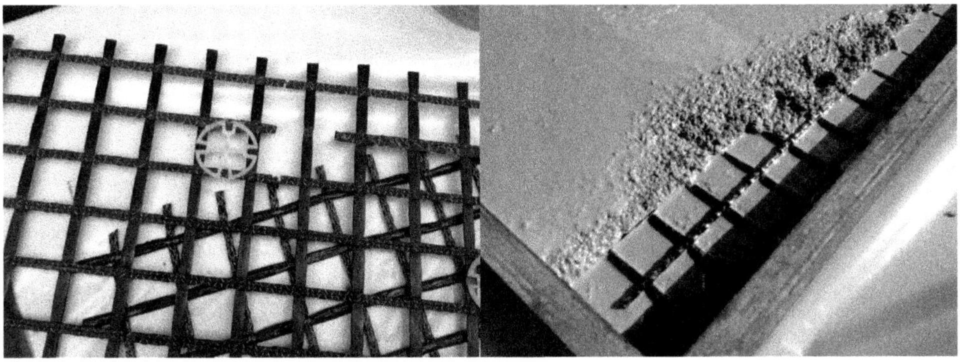

Figure 5.26 Carbon fiber in civil engineering

Table 5.3 lists some important manufacturers of fabrics and multiaxial fabrics.

Table 5.3 Manufacturers of Fabrics and Scrims and Fields of Application

Manufacturer	Materials	Application
SGL Group	fabric, multiaxial fabrics	automotive, wind energy
Saertex GmbH & Co KG	multiaxial fabrics	transportation, wind energy, aviation, shipbuilding
Hexcel Composites	fabric and spreading fabric, multiaxial fabrics	aviation, automotive, wind energy, sports
Dow AKSA	fabric, multiaxial fabrics	transportation, wind energy, civil engineering
ECC Fabrics	fabric and spreading fabric, multiaxial fabrics	aviation, automotive, wind energy, construction, transportation, medical technology
Oxeon	spreading fabric	transportation, aviation, automotive, sports

Dry Preforming and Impregnation

In the next processing step, the semi-finished products described above, such as fabric, multiaxial scrims (NCF), and dry fiber tapes (DFP), can be preformed to a final component geometry, the preform, using preform technology (Figure 5.27).

For this purpose, dry textiles are coated on the surface with a binder based on epoxy or thermoplastic (powder coating). The textile blanks are then draped (inserted) into a shaping tool and consolidated under pressure and heat.

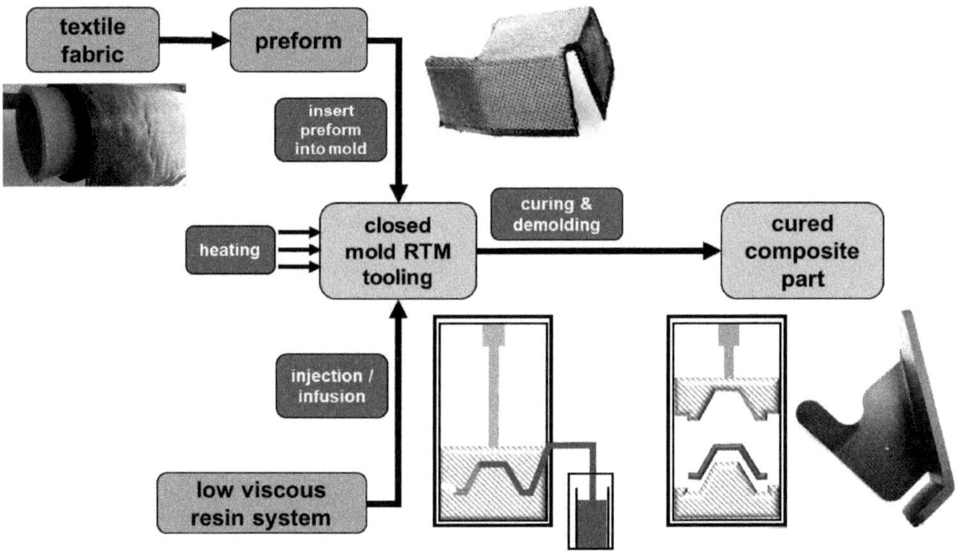

Figure 5.27 Processing of textile semi-finished products into finished components

The advantages of this preform are good dimensional stability and thus considerably easier handling than with dry textile blanks. For impregnation with a matrix system in the next step, shorter process times can be achieved, e.g. when inserting the preform into the injection or infusion mold. Preforms can be prepared and temporarily stored for series production according to demand and production rate. Figure 5.28 shows an angled CFRP assembly element (so-called case corner), which is first brought into the desired contour from bound dry fabric in the preform process and stabilized. In the second step, the preform is impregnated with matrix resin in a suitable injection mold and cured.

Figure 5.28 Preform (left) and fully cured component (right)

This method of preforming is mainly used in series production with large quantities, e. g. in the automotive industry for body parts of the BMW i3. The basic structure shown in Figure 5.29 is constructed from dry multiaxial scrim as a preform, soaked in an injection mold, and cured. The final component contour is achieved using milling techniques.

Figure 5.29 Automotive structure of CFRP based on multiaxial fabrics

5.2.2.2 Carbon Fibers, Cut or Ground

Cut and ground carbon fibers are produced from continuous carbon fibers by precise cutting to the desired length or by grinding the continuous fibers. Like the continuous carbon fibers, the short fibers can also be produced with different sizing in order to achieve optimum bonding of the fibers to the matrix. The fiber flours as fine powders are usually produced without any sizing (see also Figure 5.30). With their good mechanical properties, both configurations, short fiber and fiber flour, offer ideal conditions for a wide variety of mixing and compounding processes, especially for the mechanical reinforcement of thermoplastics that are further processed by injection molding (see also Figure 5.1). Some of the essential properties of short fibers and fiber flour are shown in Table 5.4.

Figure 5.30 Cut (left) and ground (right) carbon fibers [*Tejin Carbon*]

Table 5.4 Properties of Cut/Ground Carbon Fibers

Important Properties
▪ Excellent mechanical properties in combination with matrix systems
▪ Low density
▪ Low thermal expansion
▪ Very good electrical conductivity or shielding
▪ Non-corrosive, highly abrasive
▪ Good dispersibillty in thermoplastics and thermosets
▪ Very good pourability

Depending on the thermoplastic used, the typical fill levels are up to 30% carbon fiber content. Due to their good electrical conductivity, they are also suitable for making plastics (thermoplastics, thermosets) as well as other materials, e.g. special papers, conductive. Such materials can be used, for example, as protection against electromagnetic interference (EMI shielding), but are considerably lighter than conventional metallic shields made of copper or aluminum. Further examples of applications of carbon fibers in cut or ground form are summarized in Table 5.5.

Table 5.5 Examples for the Use of Cut and Ground Carbon Fibers

Cut Fiber	Ground Fiber
Printer and electronic components	Floor coverings and coatings
Components of fuel cells	Crude oil offshore drilling
Cement reinforcements	CNG pressure tanks
Gear wheels and mechanical components for mechanical engineering and automotive industry	3D print filament

Thus, the applications for short fibers and fiber flour are extremely diverse. Accordingly, numerous different variants are also available with regard to the sizing application. This ensures good bonding or miscibility of the fiber material with the surrounding matrix for optimum end properties.

Figure 5.31 LFT granules [*SGL Technologies GmbH*]

The classic application of fiber flours and short fibers, bound in the form of LFT granulates (see Figure 5.31), are mainly injection-molded products such as plastic housings for laptops or printer housings (Figure 5.32). In the latter application, the focus is not only on improving the mechanical properties of the material, but also on improving the electrical conductivity to dissipate static electricity.

Figure 5.32 Printer components in injection molding technology with short carbon fiber reinforcement [*SGL Technologies GmbH*]

Figure 5.33 shows an injection-molded fitting for an aircraft door for the Airbus A350-900. The fiber composite structure of the aircraft door uses a CFRP outer skin and a stiffening structure on the inside. The fitting selectively connects the outer skin of the aircraft door with the inner stiffening structure so that the two components form a box supporting structure and exploit the maximum moment of inertia of the surface. The component is manufactured from highly rigid, short carbon fiber reinforced polyether ether ketone (PEEK, VICTREX™ PEEK 90HMF40).

Figure 5.33 Fastening clip for aircraft door made of short carbon fiber-reinforced PEEK using injection molding technology [*Airbus Helicopters GmbH*]

The series solution developed by *Airbus Helicopters*, which has resulted in weight and cost reductions of 40%, replaces an identical aluminum fastening clip that has to be produced from an aluminum block by costly milling. The fiber-reinforced high-temperature thermoplastic resists continuously occurring moisture in the area of aircraft doors without any problems, whereas special surface treatment is necessary for aluminum in order to prevent corrosion. The know-how from this first application of fiber-reinforced PEEK will probably also be transferable to other components and other applications with similar requirements. The VICTREX PEEK 90HMF40 used has up to 100 times longer fatigue life and up to 20% higher specific strength and stiffness compared to 7075-T6 aluminum under the same conditions [5-13].

With the increasing use of 3D printing technology, fiber flour or short fibers are used in the corresponding printing filaments which are offered by numerous companies. Figure 5.34, right, shows the air duct for a drone propeller printed with a carbon fiber reinforced polyamide 6 (PA6/CF15%).

Figure 5.34 Left: polyamide/CF-printing filament, Fiberthree CarboLine SX.09, right: 3D-printed air duct of a drone (fan duct)

The component has a diameter of around 12 cm and was produced in just under three hours. In addition to the fast and simple production technology, the material also offers high tensile strengths of > 100 MPa in the x and y directions.

Another example of the changing world of carbon fiber manufacturing and use is the Shelby Cobra printed by Oak Ridge National Laboratory (ORNL) - a 3D-printed car. The vehicle was printed in the Manufacturing Demonstration Facility (MDF) of the Department of Energy at ORNL with the BAAM machine (Big Area Additive Manufacturing). The BAAM, developed by Cincinnati Incorporated and ORNL, is the machine with which the world's first 3D-printed car "Strati" was printed. ORNL's 3D-printed Shelby Cobra, along with many other innovations, was exhibited at the North American International Auto Show in Detroit from January 12-15, 2015. The following photos (Figure 5.35, Figure 5.36, and Figure 5.37) are taken from the corresponding press release of the ORNL.

Figure 5.35
Pressure principle of BAAM (Big Area Additive Manufacturing)

Figure 5.36
3D printing of the Shelby Cobra on ORNL's MDF

Figure 5.37
3D-printed parts of the Shelby Cobra

For several years the suitability of short fibers as reinforcing fibers in screed concrete for the construction industry has been investigated. Suitable hydrophilic adhesion sizing is required for this, which enables good miscibility and bonding of fibers with the matrix for optimum force transmission, as the fibers cannot form an adequate bond with the concrete matrix due to their inert reaction and smooth surface structure, so that there is no increase in tensile strength under load. As with classic fiber additives (polymer or steel fibers), only a small proportion of the carbon fibers should be added. The advantage over steel fibers in particular is that the carbon fibers in the alkaline screed concrete matrix do not corrode due to penetrating moisture and their load-bearing effect and crack-bridging function are not impaired.

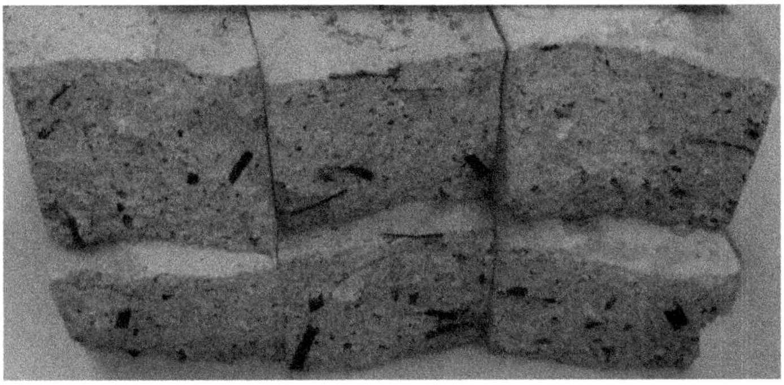

Figure 5.38 Samples of screed concrete with short carbon fiber [*CarboNXT GmbH*]

Short carbon fibers can also be further processed into very thin and extremely light carbon fiber fleeces, the basis weight of which can be varied from 2 g/m^2 to 400 g/m^2. These can be used in many different ways:

- Improvement (smoothing) of the surface of CFRP components (e.g. produced by pultrusion)
- Manufacture of ultra-thin electrical heating mats for use in construction. The excellent electrical conductivity makes the carbon fiber fleece-based heating mats an interesting alternative to classic heating systems, which can also be retrofitted with relatively little effort.
- Shielding against electromagnetic radiation (as metallized fibers)
- Carrier material for technical adhesive films (instead of polyester)

The mechanical properties are not always the sole focus, as is the case, for example, with applications in the field of offshore triple drilling. The drill pipe, which reaches to a depth of 3000 m, is coated with floating bodies for stabilization. In order to reduce the specific density of the float material and thus further improve buoyancy, hollow spheres containing carbon fiber flour are introduced into the foamed plastic.

Figure 5.39 Left: carbon fiber fleece 20 g/m², *CarboNXT,* right: surface heating based on carbon fiber fleece, Polymer Composite Consulting H. Lengsfeld

Table 5.6 shows an overview of some manufacturers of short fiber types and fiber flour.

Table 5.6 Overview of Selected Manufacturers of Short Fibers, Fiber Flour, and Nonwovens

Manufacturer	Name	Fiber Type Used
Toho Tenax Europe	TENAX®-J TENAX®-A	short fiber/fiber flour with thermoplastic, thermoset, or water-soluble preparation application
SGL Group	SIGRAFIL®	HS short fiber or fiber flour with thermoplastic, thermoset, or water-soluble sizing application, dry short fiber fleeces based on new fibers
Hexcel Composites	HexTow® chopped fiber	HS or IM short fiber or fiber flour with thermoplastic or thermoset preparation application
ZOLTEK Corporation	ZOLTEK™ PX35 chopped and milled fiber	HS short fiber or fiber flour with thermoplastic, thermoset, or water-soluble preparation application
Dow AKSA	AC types	HS short fiber with thermoplastic or thermoset preparation application
TFP	Optiveil® Optimat®	dry short fiber fleeces
CarboNXT	CarboNXT® veil CarboNXT® fiberball	dry short fiber fleeces, fiber flour, short fibers with thermoplastic, thermoset, or water-soluble preparation application
TENOVO	n.a.	dry short fiber fleeces

5.3 Pre-Impregnated Semi-Finished Fiber Products

Many dry semi-finished fiber products such as UD rovings, fabrics, and scrims can be pre-impregnated using suitable matrix systems (thermoplastics, thermoplastics) so that the material can be cured (thermosets) or consolidated (thermoplastics) directly into the finished component after being cut to size and deposited in a mold. Semi-finished fiber products impregnated with reaction resins (thermosets) are generally referred to as prepregs. Due to the large number of textile structures (UD, fabric, NCF, etc.), the different variants are often named even more precisely. Continuous fibers are known as UD prepreg, whereas impregnated short fiber semi-finished products are referred to as BMC and SMC. Impregnated, isotropic nonwovens are also offered as a further textile structure. Thermoplastic impregnated fiber structures are often called organotapes or -sheets (based on UD fibers and fabric or scrim structures).

Table 5.7 Selection of Some Manufacturers of Carbon Fiber-Based Prepregs

Manufacturer	Name	Prepreg Type	UD	Weave	NCF	C-SMC
SGL Group	SIGRAPREG®	thermoset	X	X		
Hexcel Composites	HexPly®	thermoset	X	X	X	X
Toray Industries	TORAYCA™ Prepregs	thermoset	X	X		
CYTEC Solvay Group	CYCOM®	thermoset	X	X		
Gurit	Performance Prepreg & SPRINT™	thermoset	X	X	X	
Mitsubishi	n.a.	thermoset	X	X		X
Ten Cate	Cetex®	thermoplastic, thermoset	X	X		X
Toho Tenax	TPUD, TPWF	thermoplastic	X	X		
Suprem SA	Suprem™	thermoplastic	X	X		
Bond Laminates	Tepex®	thermoplastic		X		
ARKEMA	Polystrand® Kepstan®	thermoplastic	X	X		
ECC Fabrics	ECC ePreg®	thermoplastic		X		
Continental Structural Plastics	n.a.	thermoset				X

Figure 5.40 shows examples of the different pre-impregnated variants.

Figure 5.40 Examples of impregnated semi-finished thermosets:
(1) ATL UD prepreg, (2) AFP UD slit tape and (3) fabric prepreg, (4) SMC/BMC compound; thermoplastic: (5) organosheet and (6) UD organotape
Image source (4): [*CarboNXT GmbH*]; (5), (6): [*Toho Tenax Europe*]

5.3.1 Applications: UD Prepreg (Thermoset)

UD prepregs based on reactive resin systems have been used since their development in the 1960s for the production of highly stressed CFRP structures, especially in the aerospace and racing industries. Common matrix systems are epoxy resins (EP), bismaleimides (BMI), polyester and cyanate ester resins, and polyurethanes. When processed correctly, these semi-finished products stand for the very highest lightweight construction level and component quality. With the development and production of CFRP-based fuselage structures for the passenger aircraft of Airbus (A350) and Boeing (B787), the worldwide demand for these UD semi-finished products again increased significantly in the first decade of the 21st century. In addition, numerous further developments and improvements were made. The predominant quantities of UD prepregs are available either as 300 mm wide webs or as narrow slit tapes or towpregs that are processed into laminates with automatic taper layers (ATL, AFP) [5-11]. These processes are mainly used for large components in aviation (wing and fuselage shells as well as vertical fins) (Figure 5.41). Manual processing only takes place for small components or small batch sizes.

Figure 5.41 UD prepreg construction in aviation – left: demonstrator vertical stabilizer, right: fuselage shell segment

5.3.2 Applications: Fabric and Scrim Prepreg (Thermoset)

Fabric and scrim prepregs are characterized by good manageability and simple processing in manual processes. Typical applications for fabric prepregs are cover layers in sandwich components (e.g. aircraft fairing structures), monolithic components such as stiffening elements (so-called stringers, Figure 5.42, left), and force transmission elements (so-called fittings, Figure 5.42, right) in aviation. The CFRP material offers advantages over metal materials such as very high fatigue strength and corrosion resistance, which simplify the maintenance of these components and thus reduce operating costs.

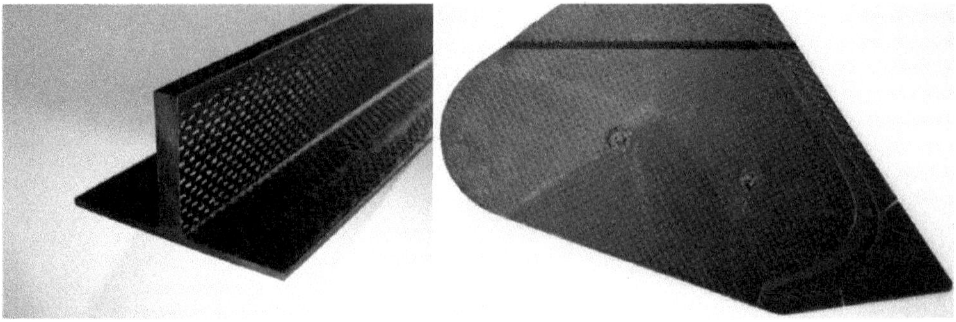

Figure 5.42 Fabric prepreg
left: stringer in pultrusion technology [*Secar technology*], right: fuselage connection fitting of an aircraft vertical tail fin

Textile fabric prepregs are used in wind and water power for the construction of rotor blades and underwater turbines. The Finnish manufacturer *Wicetec Oy* uses scrim prepregs to manufacture electric heating elements for rotor blade de-icing (Wicetec Ice Prevention Systems). The carbon fiber-based heating elements are applied to the leading edge of the rotor blades, in some cases over the entire surface. By heating of the surface, ice formation will be prevented. This prevents unfavorable flow conditions at the rotor blade (turbulent flows) and power losses of the wind turbine.

Both prepreg variants are also used in sports articles (e.g. touring skis, snowboards) and automotive components (hybrid B-pillars). For skis and snowboards the carbon fiber multiaxial fabrics in the inner laminate structure offer not only fast processability and high rigidity, but also adaptation, e.g. to the vibration damping of a ski (Figure 5.43).

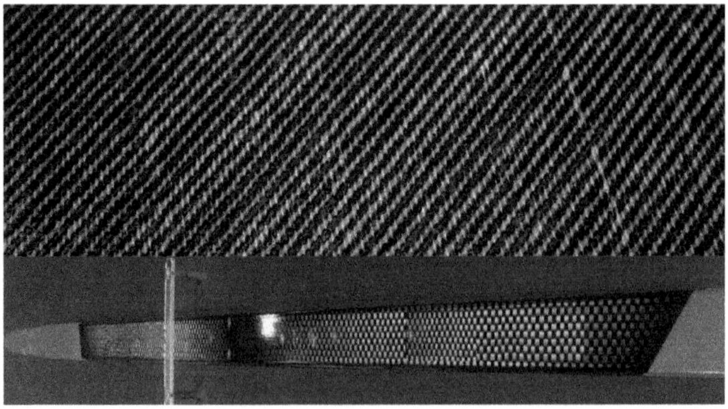

Figure 5.43 Top: triaxial NCF prepreg, bottom: application in ski sports

An example from the field of body light construction is the application of hybrid B-pillars made of e.g. high-strength aluminum with a CFRP patch based on a fabric prepreg. The aim of reinforcing such an aluminum component with highly rigid carbon fibers is to increase the buckling stability of the B-pillar in the event of a side impact so that there is no buckling in the head area of the passenger compartment. With the aid of local CFRP reinforcement, for example, a displacement of the buckling into the floor area of the vehicle can be achieved. Such a solution offers a significantly lower weight than the high-strength steels commonly used today. This technology has been in use at BMW in the 7 Series model as a steel-CFRP hybrid design since 2015.

Figure 5.44 shows a functional demonstrator, which was produced in cooperation with *BENTELER* during a material test by *Evonik* to test an innovative resin formulation for prepregs. In addition to the manufacturability of a complex, high-strength aluminum structure in a thermally supported deep-drawing process, the function of a hybrid aluminum-CFRP body component was also to be demonstrated. The special formulation of the prepreg matrix resin used permits short cycle times and is characterized by impact resistance and a durable bond between CFRP and metal, without the need to use a structural adhesive film. The hybrid design also offers the advantage that no special mold (tooling) is required for curing the fiber patch. Instead, the fast-curing prepreg patch is inserted directly into the metallic component and cured by pressing, where it bonds to the aluminum sheet in situ. The connection of the finished component with the rest of the vehicle structure in the later body-in-white process can also be carried out for each hybrid version using established joining technologies such as welding, riveting, or gluing, so that there is a high degree of flexibility in the selection of the joining technology in a later series process and thus cost savings can possibly be achieved.

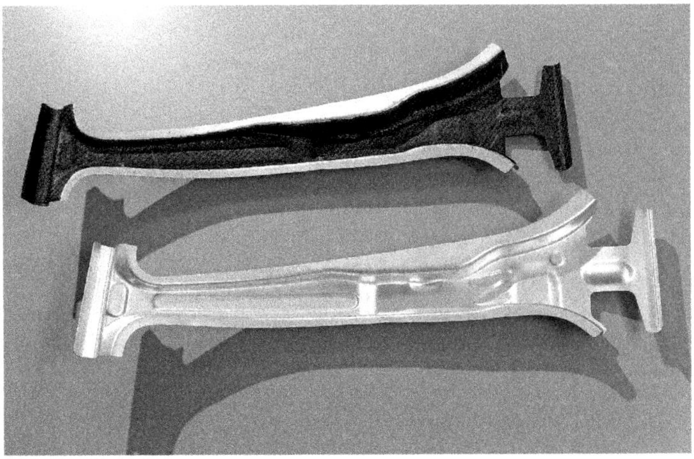

Figure 5.44 Hybrid construction – lightweight body construction using the example of a hybrid B-pillar made of high-strength aluminum with a CFRP patch [*BENTELER Automobiltechnik GmbH, © Evonik Industries AG*]

Gummiwerk Kraiburg has achieved an innovative development with its hybrid material KRAIBON®. Epoxy resin-based fabric prepregs with thin layers of non-cross-linked elastomer rubber are reactively cured together. The result is an extremely stable hybrid material made of CFRP and elastomer. By alternating the layers of hard and soft material in a laminate, the high mechanical properties of a CFRP laminate can be combined with the good elastic properties of the elastomer. This hybrid material shows excellent damping properties with regard to acoustic behavior as well as significantly improved impact and splitter behavior. These properties can be used in automotive door structures, where the elastomer component provides significantly improved acoustic damping (noise reduction of 20 dB) and at the same time low weight compared with conventional noise damping systems, so-called heavy-duty mats. The weight increase of the hybrid door structure due to the elastomer content is only 5% compared to the non-modified variant. In addition, the door structure shows improved crash behavior.

Figure 5.45 Acoustically damped CFK-HYBRID door of a KRAIBON® sports car [*Gummiwerk KRAIBURG GmbH & Co. KG*]

5.3.3 Applications: C-SMC/C-BMC

Due to their good manageability and versatile adaptability with regard to curing parameters (temperature, duration, speed), SMC/BMC molding compounds are used both for cladding and structural components in automotive, aerospace, and mechanical engineering applications. The semi-finished products can be brought into almost any geometric shape by suitable pressing processes and can be excellently reworked mechanically after curing if required. Depending on the type of semi-finished product and the process, the cured components or boards exhibit very good strength and rigidity as well as almost isotropic material behavior, unlike the typical laminate structure. The material has the potential to be used as a material in mechanical engineering alongside classical materials such as metals. Because the density is 4 times lower than that of steel, this class of materials is particularly suitable for areas with (fast) moving masses, such as robots. Due to the higher material price, the use in machines and plants has so far been limited.

Figure 5.46 Left: C-SMC plate for mechanical engineering, right: demonstrator component connection clip for aviation applications [*Marc Fette, University of Hamburg*]

C-SMC molding compounds are already being used for automotive structures such as tailgates. Here, not only individual stiffening elements such as ribs are produced from the C-SMC, but also the entire supporting structure of the liftgate.

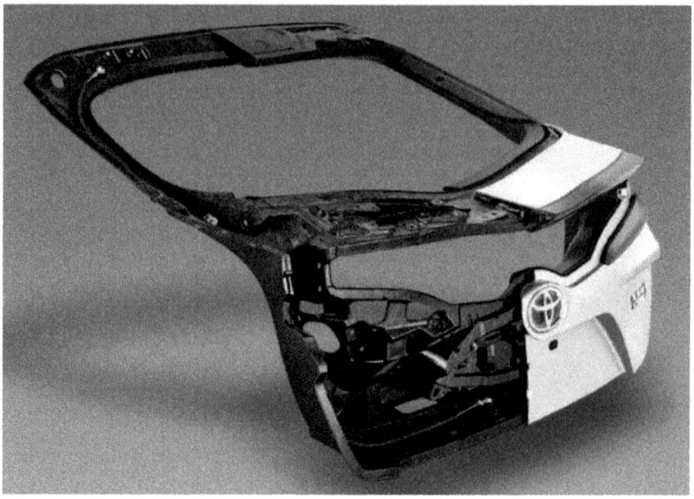

Figure 5.47 Toyota uses carbon fiber SMC from Mitsubishi Rayon for the tailgate of the Prius PHV from model year 2018 (Green Car Congress, April 24, 2017)

The SMC developed by Mitsubishi Rayon (MRC) is a kind of intermediate material in the form of a fiber mat for CFRP, in which carbon fibers cut into pieces several centimeters long are dispersed in resin. The SMC can be processed into components by compression molding in short cycle times – about 2 to 5 minutes. In contrast to prepreg intermediate materials (uncut carbon fiber fabric impregnated with resin), this SMC is characterized by high formability for forming complicated molded parts.

Carbon SMC also shows almost uniform mechanical properties. In this way, engineers can easily utilize the carbon fiber material by drawing on the know-how of existing parts to achieve lighter, stronger components.

The advantages of MRC's SMC for the Prius PHV include a significant weight reduction of the vehicle, achieving high component performance, excellent formability enabling the manufacture of complex shaped components, and the productivity required to produce components for production vehicles.

5.3.4 Applications: Organotapes

Organotapes represent a consistent further development of the classic thermoset-based prepregs, which have clear advantages over reactive prepregs, especially in terms of storage and handling. As the range of C-fibers and thermoplastics in matrix form (e.g. PA12, PES, PEEK, PPS, TPI) increases, as well as the development of the (automated) system technology required for processing, the number of applications (e.g. in the sporting goods, medical, protective clothing, and automotive sectors) has also increased. As a rule, these are still relatively small components that are frequently produced by pultrusion or AFP processes. In aviation, there are currently no large structural components made from organotapes, such as fuselage, wing, or tail components. However, changes are to be expected in this market in particular, as thermoplastic semi-finished products and components produced from them are expected to offer advantages over metallic or thermoplastic prepreg-based materials both in processing and handling and in subsequent applications.

Figure 5.48 Application – pultruded profiles and rods based on UD fibers in thermoplastic matrix [*Suprem SA*]

Usually, the organotapes are laid into laminates and consolidated into organosheets (hot presses). These semi-finished products can then be further processed in further steps, e. g. in a hot forming process, to produce the finished component. In addition, structures made from tapes (direct deposition or organosheet forming) can be encapsulated with an additional matrix in a downstream injection molding process and thus obtain the final component contour without additional mechanical processing. Figure 5.49 shows a baby bath as a demonstrator for this process combination. In the first process step, thermoplastic UD tapes were laid flat and hot formed in the forming tool. The additional, stiffening rib structure was then applied in the same tool by injection molding.

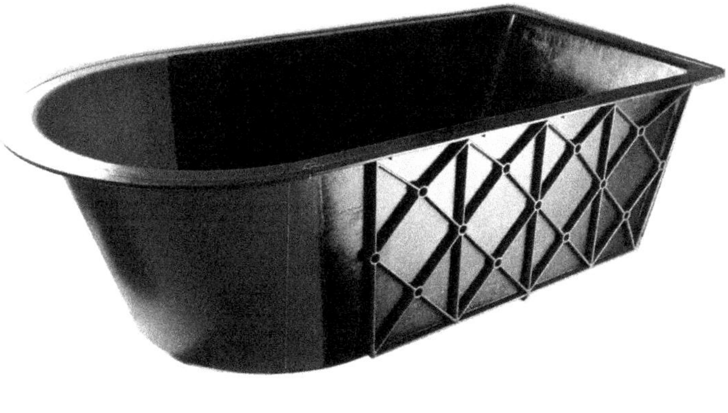

Figure 5.49 UD-Organotapes in combination with injection molding [*Neue Materialien Bayreuth GmbH*]

5.3.5 Applications: Organosheet

Organosheets (UD or fabric) are consolidated either by depositing organotapes or with thermoplastic-coated fabrics in hot pressing processes. Using the same process principle, the organosheets are then formed into the final component geometry and, if necessary, mechanically reworked, e. g. by water jet cutting.

Organosheets are used in:

- Automotive (e. g. pedals) and aerospace (assembly clips),
- Consumer electronics (e. g. mobile phones, notebooks, printers),
- Sports articles (e. g. sports shoes, bicycle components, protective clothing, helmets)
- Industry (e. g. shock absorbers),
- Aviation, in which formed organosheets are used as connecting elements (so-called clips) in fuselage assembly, e. g. for the Airbus A350 (Figure 5.50). In contrast to a classical solution with aluminum clips, these offer the advantage that no galvanic corrosion can occur between the skin fields of the CFRP fuselage and the organosheet clips. In the case of metallic clips, these must always be protected against electro-galvanic corrosion with a protective lacquer, since in this case the CFRP component is has a lower oxidation potential than the metal component.

Figure 5.51 shows a section of a printer component, for the production of which an organosheet is formed under pressure and temperature and then additionally back-injected with thermoplastic (ribbing).

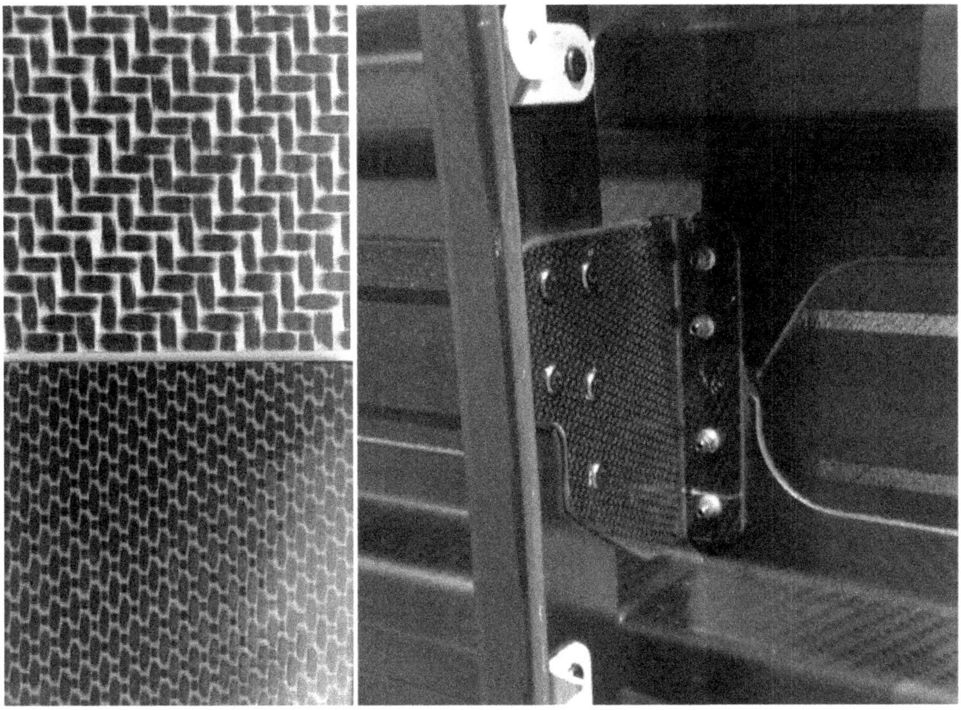

Figure 5.50 Left: organosheets based on polyether sulfone (PES) with various fabrics, right: application as connecting element in aircraft construction

Figure 5.51 Thermoformed organosheet with back-injected ribs [*SGL Technologies GmbH*]

Outlook

The numerous application examples of carbon fibers presented in Chapter 5 show how versatile this material can be used due to its properties and how diverse the variations and combinations with matrix materials are. The list of examples could be continued almost endlessly and the number of applications continues to increase. This is not least due to the fact that the price of carbon fibers has already fallen in certain areas due to new and further developments (see Chapter 7). Due to the high intrinsic value of carbon fibers and products made from them, especially in view of the energy already used, the importance of recycling production residues as well as end-of-life components (EoL components) is steadily increasing. In recent years it has become increasingly possible to use recycled fibers for high-quality applications. Chapter 6 therefore also deals with recycling.

References for Chapter 5

[5-1] Company brochure SGL Technologies GmbH – *PANOX® Oxidized PAN fibers*, status 2017

[5-2] Company brochure Toho Tenax – *PYROMEX®*, status 2017

[5-3] H. Jäger, T. Hauke: *Carbonfasern und ihre Verbundwerkstoffe*, Süddeutscher Verlag onpact GmbH, Munich, 2010, ISBN978-3-86236-001-7

[5-4] Company brochure ZOLTEK – *ZOLTEK™OX*, status 2017

[5-5] Company brochure Hexcel Composites, *HiTape® and HiMax®*, status 2017

[5-6] Company brochure Cytec Solvay Group, *PRISM® TX1100/EP2400*, status 2016

[5-7] Company brochure Hexcel Composites, *Direct Processes Technology*, status 2017

[5-8] Company brochure SEARTEX GmbH, *flyer Multiaxial fabrics*, status 2017

[5-9] Company brochure SGL Technologies GmbH – *Short fibers PAN fibers*, status 2017

[5-10] Company brochure Toho Tenax – *short fibres PAN fibres*, status 2017

[5-11] H. Lengsfeld, F. Wolff-Fabris, J. Krämer, J. Lacalle, V. Altstädt: *Composite Technology: Prepregs and Monolithic Part Fabrication Technologies*, Carl Hanser Verlag Munich, 2016, ISBN 978-1-56990-599-9

[5-12] Wicetec Ice Prevention Systems, *www.wicetec.com*, release date November 2017

[5-13] Press release VICTREX *www.victrex.com/de/news/2015/09/airbus-bracket*, accessed December 9, 2017

[5-14] Product information Gurit Services AG, *http://www.gurit.com/*, accessed November 2017

6 Recycling and Sustainability

Together with Tim Rademacker

Waste containing carbon fibers arises at various stages and in various manifestations during the production, processing, and use of carbon fibers and CFRP components made from them. This material cycle begins with the production of carbon fibers and their further processing into textile semi-finished products, where offcuts and residual spools with dry fibers have to be disposed of as waste. During further processing, e.g. into prepregs, not only dry fiber residues but also matrix-impregnated residues and possibly faulty prepreg rolls, which are scrap, arise. Finally, further waste is generated during the processing of both dry fibers and prepregs into a component as well as through mechanical finishing, e.g. milling, drilling, and grinding.

6.1 Introductory Remarks

Resource conservation and waste reduction through appropriate recycling technologies as well as the reuse of valuable materials are decisive factors for the further applicability of carbon fibers in composite structures in view of market developments regarding the production and use of these fibers (Figure 6.1, see also Chapter 3). Despite the recycling technologies that have been in existence for about 10 years and their continuous further development, the questions of sustainable reusability of end-of-life (EoL) componentsand the material used for production (CFRP) have not yet been conclusively addressed. At the same time, there is a steady increase in the production and use of carbon fiber products and thus also waste (Figure 6.2). The ability to recycle this high-quality material is today already a decisive factor in the basic material selection in many application areas, e.g. in the automotive industry with its very stringent standards.

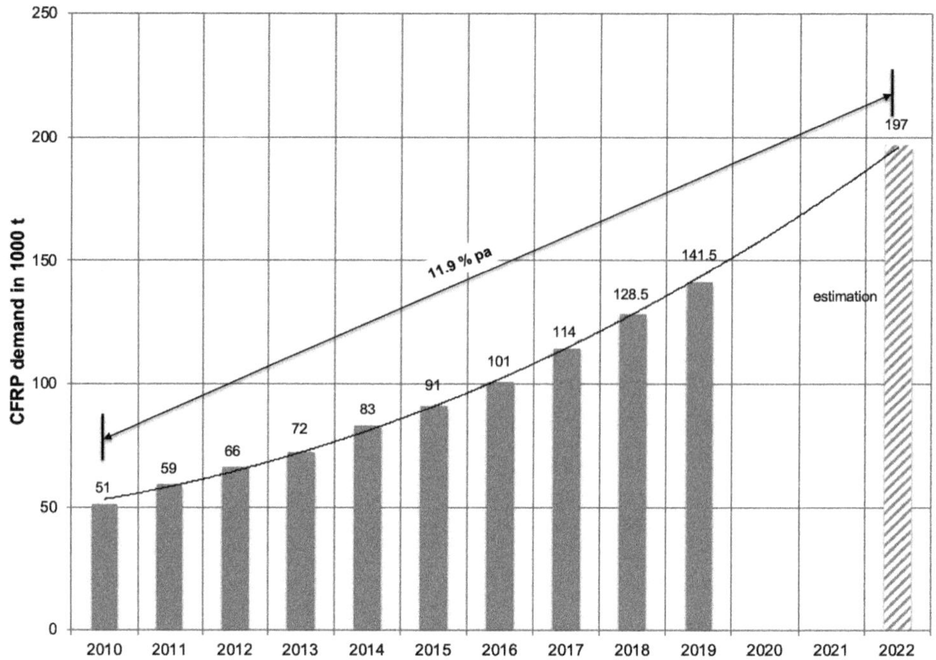

Figure 6.1 Global CFRP demand, as at 09/2019 (according to [6-1])

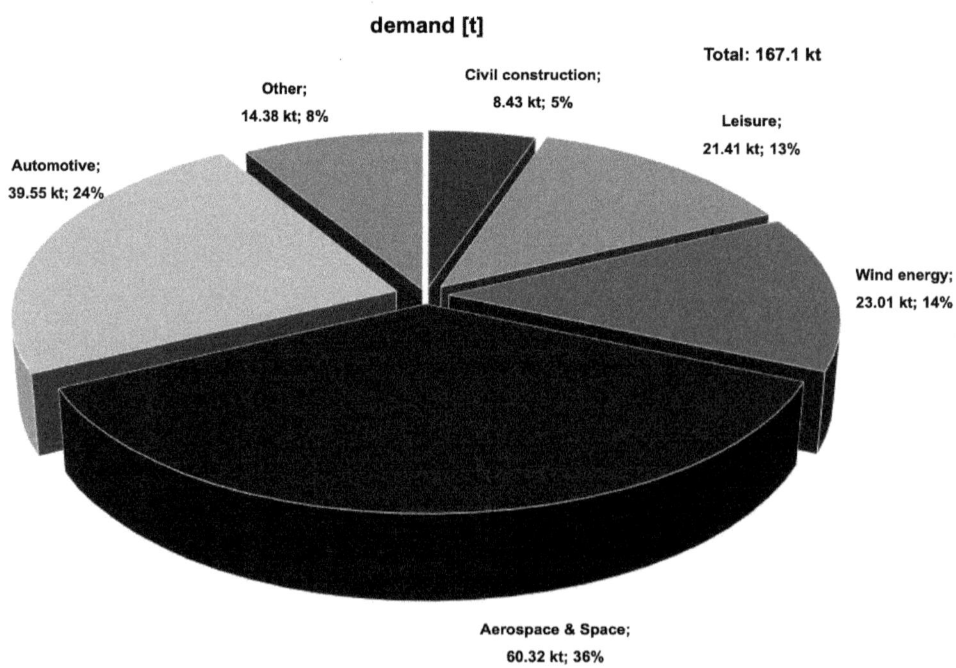

Figure 6.2 Global CFRP demand and fields of application, status 09/2019 (according to [6-1])

Carbon fiber reinforced plastics are not readily suitable for recycling due to the material combination typically used and the extremely stable combination of fiber and polymer matrix (predominantly thermoset-based matrix resins). Nevertheless, numerous studies and research projects in recent years have shown that carbon fibers and products made from them can be recycled to some extent. However, the effort involved is not insignificant. The reason for this lies in the properties designed for durability, such as mechanical strength and thermal and chemical stability.

Essentially, the recycling of carbon fibers and CFRP components made from them is determined by the factors of ecology, economy, and legal regulations. On the legal side, the EU Regulation 2008/98/EC stipulates that negative effects on the environment and health must be reduced byavoiding or reducing waste and conserving resources. The following sequence applies:

- avoidance
- preparation for reuse (component recycling)
- recycling (material)
- thermal utilization
- landfilling

Landfilling of composite waste and EoL components has been prohibited in Germany since 2005 and is therefore not an alternative to recycling. Application-related regulations such as the End-of-Life Vehicle Ordinance prescribe precise recycling quotas, which pose a major challenge for the automotive industry. A 95% reuse of materials from end-of-life vehicles must be demonstrated here. The production of one metric ton of carbon fiber (new fiber) consumes more than 200 GJ of energy and costs around € 10,000 (ca. US$ 12,000). Therefore, recycling of this very valuable material is to be regarded as meaningful under economic as well as ecological aspects despite the high expenditure. This contributes significantly to an improvement of the ecological footprint as well as to a reduction of CO_2 emissions. As a result, CF waste is increasingly seen as a recyclable material rather than waste. In a first step, a direct exchange of expensive carbon fiber (virgin material) for a recycled variant (often referred to as **rCF, recycled carbon fiber**) would be desirable from an economic point of view. However, this can only be achieved for short fibers or milled fiber, since recycled fibers are generally only obtained as short fibers with a length of approximately 100 to 150 mm in the available processes. New applications can therefore be seen above all in the area of short fibers and milled fiber, in which cut virgin material is used today. There is currently no consistent, standardized classification of recycling materials from carbon fibers. A possible proposal oriented along the manufacturing process from fiber to component could look like Figure 6.3. The above-mentioned change of perspective, in which the residual material from the various production steps is no longer regarded as waste, but as a valuable and raw material, is quite essential in the overall context of recycling.

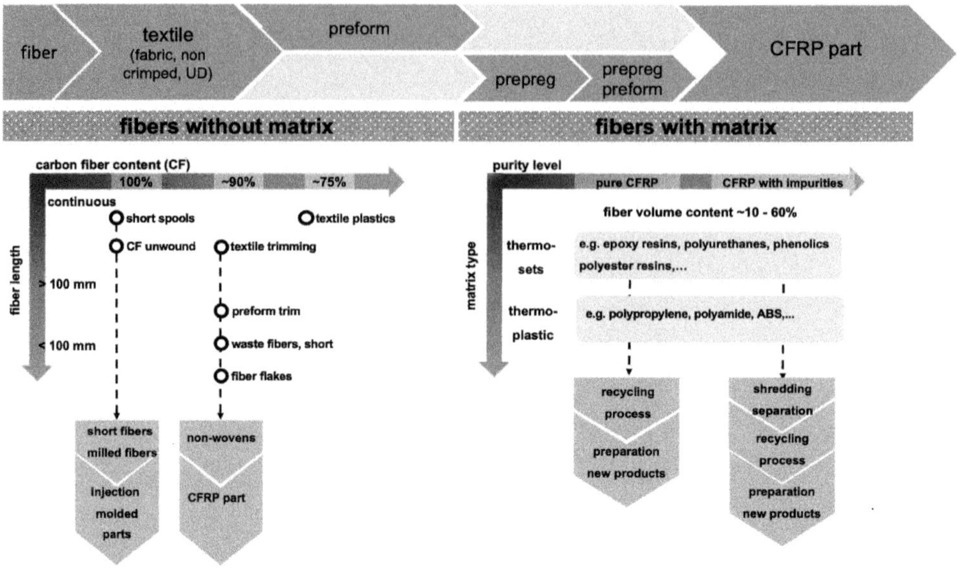

Figure 6.3 Proposal for classification of carbon fiber waste (according to [6-3])

Figure 6.3 schematically shows the processing sequence from the dry fiber to the finished CFRP component (from left to right). The procedure can be divided into the following two sections:

1. Processing of the fibers without matrix (section 1, left)
2. Processing the fibers with matrix (section 2, right)

In the first section, the carbon fibers are first pre-processed without a shaping matrix (dry). Even small quantities of binders can be used to produce preforms. Preforms are made of dry fibers which are kept in shape only by a small amount of adhesive (high-melting epoxy resins) or sewing threads for further processing, e.g. in a resin infusion process. Therefore, the carbon fiber content in the above diagram (section 1) decreases slowly from left to right (right arrow). At the same time, waste residues arise during processing, so that the fiber length also decreases (arrow pointing downwards). The diagram shows an example of the form in which the waste can be used in new applications.

Section 2, right in the diagram, focuses on fibers with matrix (thermosets: uncured or hardened, thermoplastics). Here, too, a distinction is made between two degrees of purity. With "pure CFRP", production residues can be directly recycled or reworked into new products. This applies both to a thermoplastic matrix and to uncured thermosets. For "CFRP with impurities" such as thermoset CFRP components, more complex processes must be used to separate the fibers and matrix again.

■ 6.2 Recycling Cycle and Challenges

The scheme in Figure 6.4 shows three available ways to recycle carbon fiber-based components and waste.

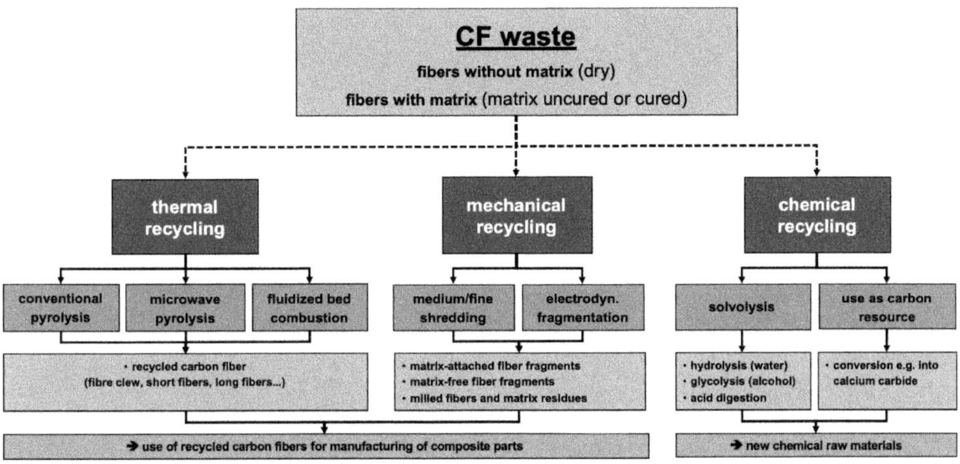

Figure 6.4 Possible process paths for recycling carbon fibers

In the thermal recycling of CFRP, the material is subjected to high temperatures to separate components. These cause decomposition of the organic matrix component (thermoplastic or thermoset matrix). As shown in Figure 6.4, the necessary temperatures can be generated using different process paths. In mechanical recycling, dry fibers as well as hardened residues are shredded into fragments of different sizes, which can then be fed back into the manufacturing process of new components (see also Figure 6.6). In chemical recycling, the carbon fibers can be recovered by solvolysis or the matrix can be preserved in the form of new chemical raw materials. On the other hand, CFRP residues can also be completely converted directly into new raw materials in chemical processes as a carbon source.

In order to achieve a positive life-cycle assessment, the aim of recycling carbon fibers must not only be a one-off recovery, but a repeated run through this process in order to achieve multiple use of the same fiber (Figure 6.5), albeit in a modified geometric form and with other applications. Starting with a continuous yarn, the use in long- and short-fiber fleeces would lead to fiber flour. Finally, it would be possible to feed the fiber-polymer flour for example into a chemical direct conversion of carbon for the production of new chemical raw materials.

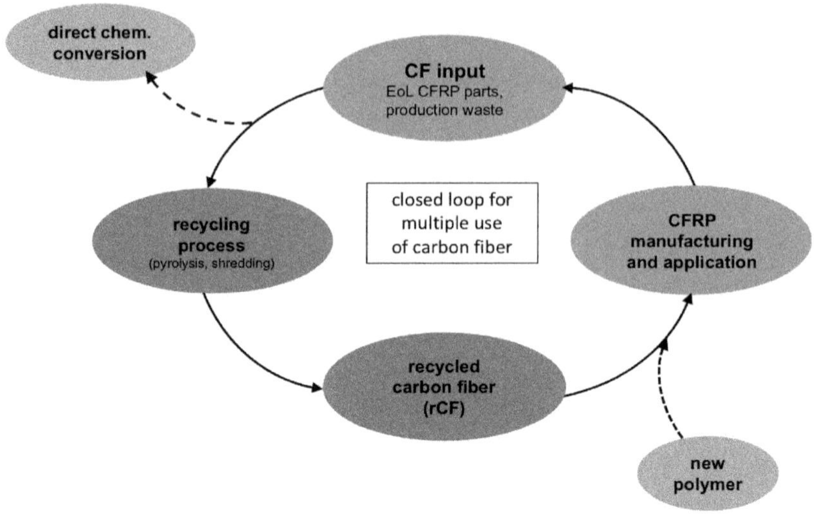

Figure 6.5 Schematic of a "closed-loop" concept for the multiple use of carbon fibers

While the closed cycle for thermoset-based carbon fiber products (cured or uncured) has not yet been implemented and is probably more complex, this principle has already been implemented for thermoplastic-based CFRP products and is offered in the form of corresponding rCF products, e. g. from *Tejin Carbon*, Wuppertal, Germany (Figure 6.6).

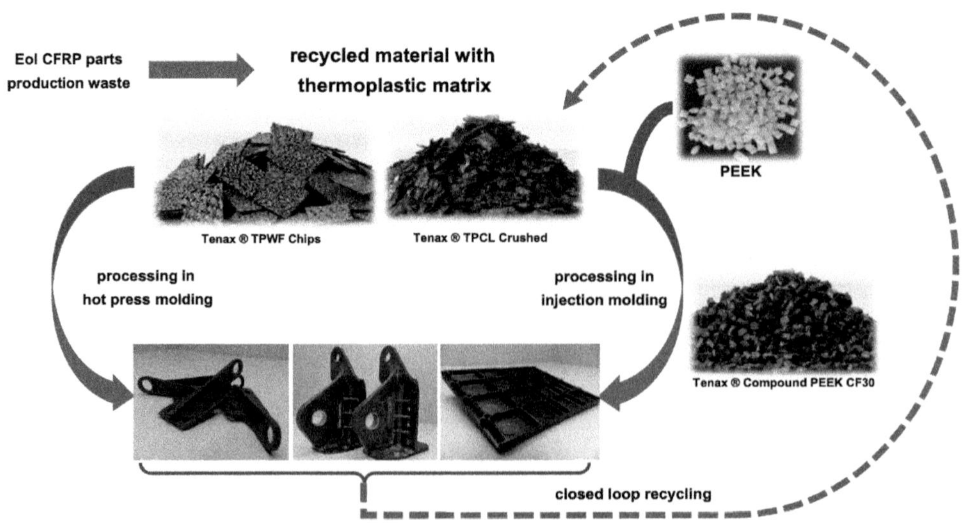

Figure 6.6 Closed-loop concept for thermoplastic-based carbon fiber products (*Tejin Carbon Europe GmbH,* [6-6])

In contrast to carbon fiber-based products and components, the recycling of glass fiber composites (GRP) by separating the fiber from the matrix is not yet worthwhile due to the much lower price of glass fibers. Either the thermal utilization and use of ash containing silicon dioxide (e.g. for cement production) or the use in crushed form as reinforcing material in cast concrete can be seen here.

6.2.1 Process and Plant Challenges

On the way to a recycled fiber, carbon fibers present some challenges due to their properties and appearance. Additional requirements for the machining processes and equipment arise from the desired properties of recycled fibers in order to be able to use them in new applications. The carbon fibers obtained from the above recycling processes (Figure 6.4) are, with the exception of dry fiber spools, generally not continuous filaments and typically have a wide fiber length distribution. When UD laminates or prepregs are recycled, quasi-continuous fibers are obtained, but they cannot easily be reprocessed as such into a new semi-finished product for composite production. These fibers are often in the form of very voluminous fiber balls with an unstructured, coarse fiber architecture ("nest formation"). Due to the large volume, very low densities of approximately 5 kg/m^3 are obtained. The fiber nests must be opened for further processing so that the fibers can be cut, aligned, or ground. In the case of pyrolyzed textile structures such as fabrics or felts, these must also be separated before further processing and the fibers separated as far as possible. No large forces must be applied, as the rCFs react very sensitively to mechanical stresses and break. Due to the fineness and lightness of the filaments ("angel hair", Figure 6.7, right), impurities (e.g. foreign fibers, matrix residues, pyrolysis coke, fine particles) can only be separated to a limited extent by mechanical processes.

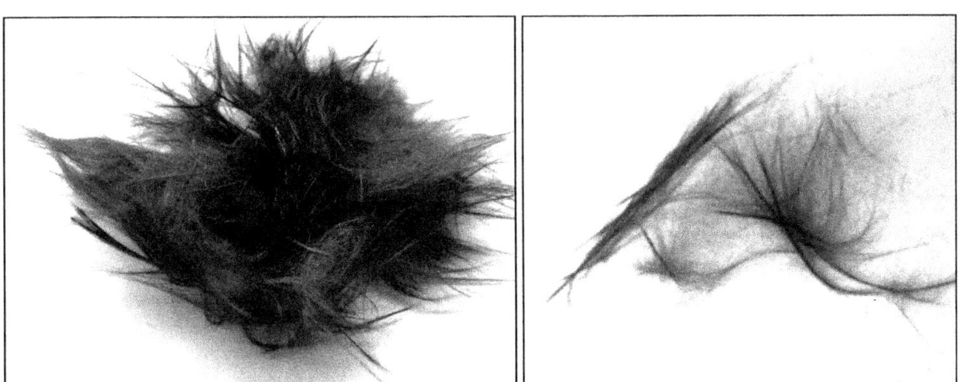

Figure 6.7 Left: rC fiber ball; right: angel hair

The excellent electrical conductivity requires special protection of electrical and electronic plant components during fiber preparation. At the same time, this property makes it difficult to separate from other materials, e.g. electrostatic particle filters. Last but not least, due to possible health hazards caused by fiber dusts, adequate occupational safety must be ensured.

The aim is therefore to obtain recycled fibers with the following essential properties:

- Specific fiber length or controlled fiber length distribution
- Defect-free fiber surface
- Maximum mechanical properties
- Fiber free of foreign matter, matrix impurities, or pyrolysis coke
- Dosable and transportable

Table 6.1 compares some properties of new fibers and recycled fibers (no short spools from the manufacturer or from roving processing), which illustrate the challenges for a later use of recycled fibers.

Table 6.1 Differences and Similarities between Virgin Fibers and Recycled Fibers

Requirement Criterion	New Fibers	Recycled Fibers
Defined carbon fiber type	yes	possible with separate collection, separation, and processing
Production of milled fiber	yes	yes
Production of cut fibers	defined length possible	more or less wide length distribution
Continuous filaments/rovings	yes	staple fiber rovings
Manufacture of fabrics	yes	small patches of staple fiber rovings, if necessary
Possible impurities	none	e.g. glass fibers, pyrolysis coke, matrix residues

The above-mentioned requirements thus clearly show how decisive both the fiber disintegration and the fiber preparation are for further processing and reuse.

6.2.2 The Challenge of Fiber Quality

The great challenge in recycling carbon fibers is to obtain a secondary fiber (rCF) of high (min. 95% level of new fibers) and consistent quality. The quality level determines the possible applications of rCF, e.g. in hybrid yarns, injection molding applications, or nonwovens. The aim is therefore to obtain recycled fibers with the following essential properties:

- Mechanical properties rCF at 95% output fiber level
- Specific fiber length or controlled fiber length distribution
- Defect-free fiber surface, no impurities
- fiber free of matrix impurityor pyrolysis coke
- Appropriate preparation application (sizing)

Table 6.2 compares some properties of new fibers and recycled fibers (no short spools from the manufacturer or roving processing) which illustrate the challenges of rCF for reuse.

Table 6.2 Differences and Similarities between Virgin Fibers and Recycled Fibers

Requirement Criterion	New Fibers	Recycled Fibers
Defined carbon fiber type	yes	possible with separate collection, separation, and processing
Defined geometric shape of the fiber, no "nest formation"	given due to the manufacturing process	mostly unstructured, coarse fiber architecture in the form of fiber balls
Defect-free fiber surface	yes	surface damage/holes possible
Production of milled fiber	yes	yes
Production of cut fibers, defined length necessary	defined length possible	more or less wide length distribution
Continuous filaments/rovings	yes	staple fiber rovings
Manufacture of fabrics	yes	small patches of staple fiber rovings, if necessary
Possible impurities or residues	none	e. g. glass fibers, pyrolysis coke, matrix residues
Good fiber-matrix adhesion	given by preparation application	given by preparation application

In the following, some methods for determining the parameters mentioned in the text or Table 6.1 are briefly described. As these methods relate specifically to rCF, they are not included in Chapter 4.

Fiber Measurement and Fiber Length Determination

After the recycling process, rC fibers usually do not have a spherical form, but rather a bulb form, so that classical methods of particle analysis, such as sieve analysis, cannot be used. In this case an image-analytical measurement can be used. In order to be able to be captured by image analysis, the rC fibers must be present in isolated cases, i. e. a measurement should possibly only be carried out after a sufficient opening of a fiber nodule. Due to the filament fineness (d = 7 to 8 µm), a high-resolution camera is required for fiber detection. In order to avoid falsification of the result, cross-over fibers, fiber branches, and fiber curvatures (geodetic length) must also be detected. Furthermore, a sufficiently large sample quantity is necessary to obtain a representative result. The problem of existing fiber measurement systems is the low throughput (static image acquisition), which hardly allows a representative statement. Dynamic image analysis systems are better suited. Last but not least, the analysis system (camera as well as evaluation software) must be able to deal with the multiscale problem, since the length of the rC fibers is in the millimeter (mm) to centimeter (cm) range, but their width is in the micrometer (µm) range.

Figure 6.8 shows a typical fiber ball after the recycling process and the corresponding microscopic black-and-white image.

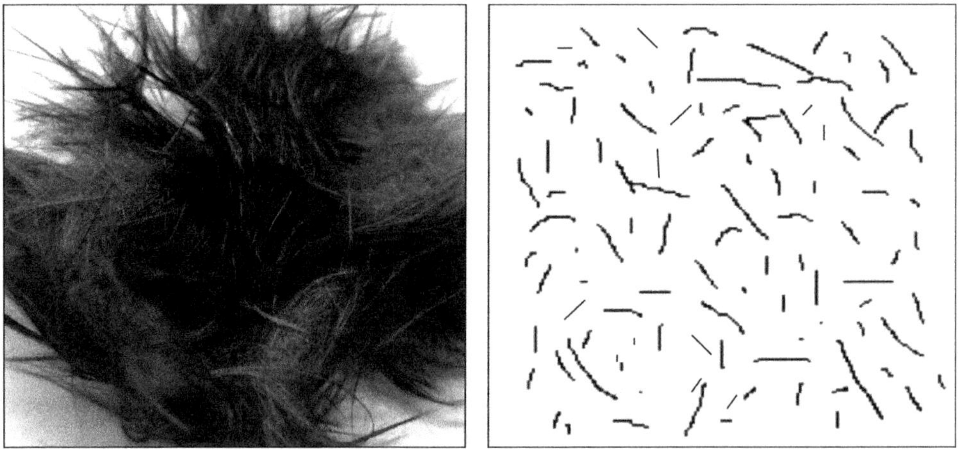

Figure 6.8 Left: fiber ball; right: microscopic image for fiber length determination

Table 6.3 Fiber Measurement and Fiber Length Determination

- High-resolution camera required for fiber detection (d = 7.0–8.0 µm)
- Multiscale problem (length in mm–cm range, width in µm range)
- Detection of cross-over fibers and fiber branches
- Observation of fiber curvatures (geodetic length)
- Sufficiently large sample volume required (dynamic image analysis)
- Problem of existing fiber measurement systems: low throughputs (static image acquisition) hardly allow representative statements

Determination of Fiber Purity

As already described above, fiber purity is another important quality parameter for the further applications of secondary fibers (rCF). Here it is necessary to determine the proportion of pyrolysis coke or matrix residues on the fiber surface. In principle, this quantitative determination can be carried out by chemical or thermal elimination of the plastic matrix component. However, the challenges for an industrial implementation are manifold. With both methods, only a small sample quantity can be analyzed and the samples cannot be tested non-destructively. Furthermore, the processes do not allow inline control during production. Another possibility would be the image analytical determination by scanning electron microscope (SEM) (Figure 6.9), which, however, makes only a very small sample section visible and is also very complex and expensive. There is therefore still a need for development in the sense of automated testing technology or process-stable recycling processes that allow a representative statement to be made with the use of the analysis techniques mentioned above.

Figure 6.9
SEM image of a filament made of recycled carbon fiber [*CarboNXT GmbH*]

Determination of the Surface Damage of the Fiber and Preparation Application

Recycling processes such as pyrolysis can cause damage to the fiber surface, which can be detected by scanning electron microscopy (SEM) (Figure 6.10) or high-resolution microscope images. Here, too, only a small selected sample section can be considered, which hardly permits a representative statement, let alone a quantitative statement on the damage. In addition, there is again a high preparation and execution effort.

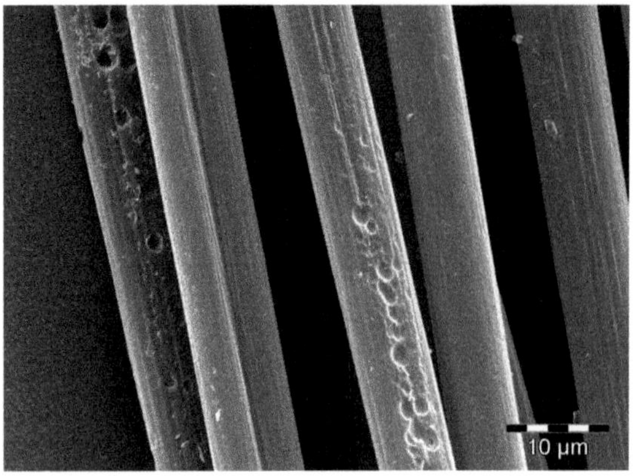

Figure 6.10
SEM image rCF with holes in the filament surface [*CarboNXT GmbH*]

In addition to possible surface damage to the fiber, it may be necessary, depending on the recycling process, to check whether and, if so, how much preparation (sizing, see Chapter 3) is applied to the fiber surface. The preparation application is a matrix-based protective layer that is applied to the fiber immediately after the manufacturing process to achieve improved fiber-matrix adhesion. The check for the existence or quantity of the sizing may be necessary, for example, if it is unclear whether the preparation order has been completely removed by the recycling process. However, due to the chemical similarity it is very difficult to differentiate between matrix residues and preparation. Apart from a few exceptions, all carbon fiber types (roving form) are usually provided with a sizing during the manufacturing process. For a subsequent application of the sizing, e.g. on a rCF fleece, a quantitative determination may be important. Since this coating is extremely thin (nanometer range) on the fiber surface and thus hardly detectable by image analysis, a thermogravimetric test (Figure 6.11), for example, is a good choice.

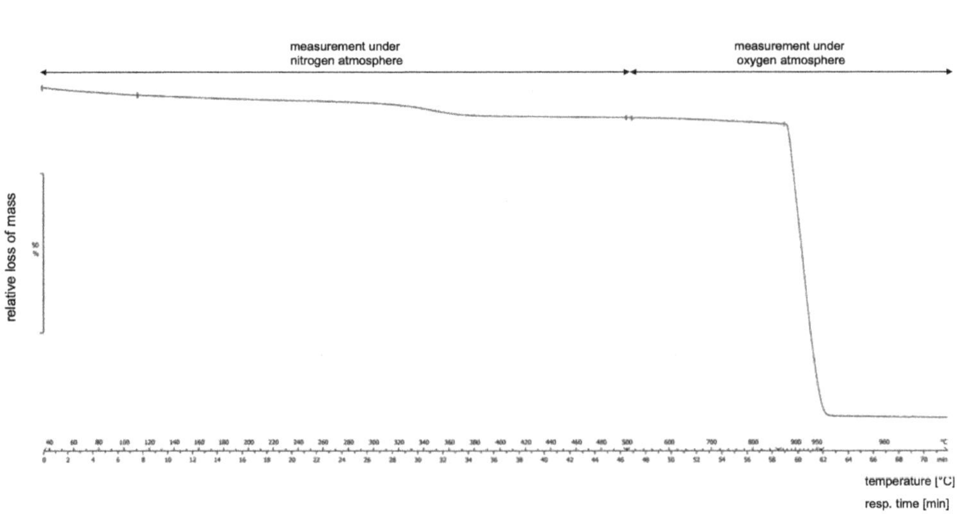

Figure 6.11 Thermogravimetric measurement of rC fibers Determination of Mechanical Properties

To determine the tensile strength by means of a single fiber tensile test, a certain fiber length is required to insert the single fiber into the clamping device (cf. Chapter 4). Here a representative, statistical statement is critical due to the sample quantity. This can be improved by testing the rC fibers in the composite, but is still difficult and time-consuming. Since there are hardly any or no comparable characteristic values of primary fibers available so far, a representative statement or a comparison with test results of secondary fibers is not possible.

6.3 The Challenge of Waste Treatment

The path to the reuse or recycling of carbon fibers usually begins with the provision of the CFRP waste by the processors. A "colorful" mix of the most diverse materials and in the most diverse manifestations is often delivered to the recycling companies (Figure 6.12).

Figure 6.12 Material mix of various residual materials from CFRP production [*CarboNXT GmbH*]

Such a material mix poses a clear challenge on the way to recycled carbon fibers. In Figure 6.13 a schematic process sequence is shown using the example of pyrolytic recycling.

Figure 6.13 Schematic process sequence using pyrolytic recycling as an example

Before the fiber material can be fed into a recycling process at all, it must first be acquired and checked for its composition. If necessary, the material is pre-cut into manageable sizes or is already available in a suitable size. This process step is of particular importance because, depending on the desired further processing, the type of pre-treatment has a decisive influence on the later area of application of the recycled material due to the resulting fiber lengths.

Preliminary disassembly is advantageous for large components. This leads to a reduction in volume and significantly facilitates the handling of flat parts. The material combinations presented in Table 6.4 may make shredding more difficult. There is still a need to optimize the existing shredding processes with regard to wear and throughput of the plants. A simplification could conceivably include appropriate separate collection and integration of CFRP waste into existing recycling structures for post-consumer flows (e. g. electrical, sports, and leisure products). A further fine shredding and processing of the waste as bulk material makes it easier to handle and produces a uniform particle size distribution and specific surface area. The maximum fiber lengths depend on the grain size of the bulk material. At the same time, fine shredding to bulk material means higher wear on the shredding systems and limits the throughput. In contrast, fiber disintegration, i. e. the separation of fiber and matrix, from coarser material may become more difficult or more complex. However, cutting the rCF fibers after they have been exposed means less wear on the cutting systems than with matrix-containing material.

The valuable fiber material (CFRP and dry fibers) is freed as far as possible from foreign substances (Table 6.4), as these impede further processing or have an unfavorable effect on the quality of the recycled fiber. Essentially, the separation is carried out by manual sorting. Furthermore, CF waste should be separated according to type wherever possible. Varietal purity in this context means that a distinction must be made between the two large groups of fiber types: HT fiber and IM fiber. This is necessary in order to achieve properties that are as constant as possible for later applications and is therefore an important quality feature. After this step, a shredding of the fiber material, the actual exposure (in the case of CFRP) of the carbon fibers, can be started (see Section 6.4). This is usually followed by processing steps to produce new semi-finished fiber products or other products.

Table 6.4 Material Mix as Input Variable for CFRP Recycling

Material	Outward Form
Without matrix (dry fibers)	roving spools weave multiaxial fabrics non-wovens
With matrix (thermoset resin uncured)	prepreg waste UD tapes fabrics and multiaxial fabrics

Material	Outward Form
With matrix (thermoset resins cured or thermoplastic)	CFRP components (end-of-life) organotapes and sheets residues from mechanical processing
Impurities	glass and polyester threads aramid fibers GFRP components metals (e. g. steel, titanium, copper) plastic films paper webs and cardboard tubes wood aluminum and phenolic resin honeycombs

Compared to the material mix shown above, it is much easier to prepare dry waste products such as fabrics, short spools, and preforms directly. After the usual collection, the dry fibers can be crushed and opened so that alignment and subsequent bonding, e. g. to nonwovens, is possible (Figure 6.14).

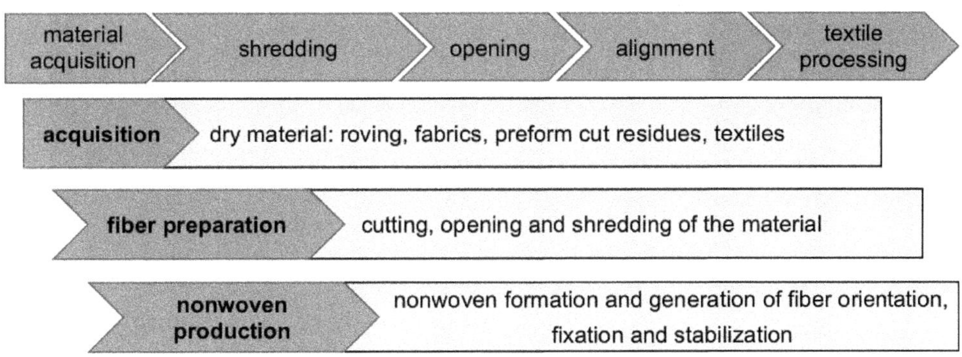

Figure 6.14 Schematic sequence for the recycling of dry fiberwaste (after [6-3])

■ 6.4 Recycling Processes for Carbon Fibers and CFRP

Figure 6.15 shows a simplified overview of possible ways to reuse carbon fiber containing residues or end-of-life components. In principle, there are three groups of procedures available, which have already been mentioned above:

- Thermal processes

- Chemical processes
- Mechanical processes

While some processes only recover fiber material, other techniques can also recover part of the organic matrix polymer in a different form. The use of CFRP residues as a carbon source even offers the possibility of complete material-thermal recycling and conversion into new materials. On the other hand, purely thermal recycling offers only a one-off energy gain from the calorific value contained in the material.

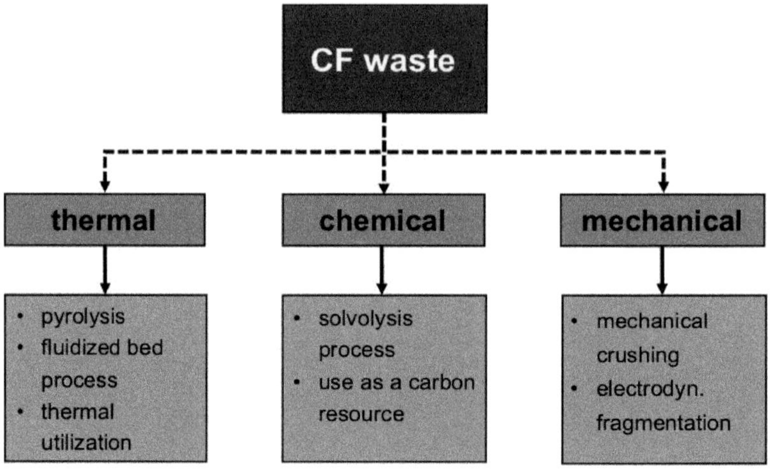

Figure 6.15 Ways to recycle CFRP

6.4.1 Mechanical Shredding

In mechanical recycling, cured CFRP composite material or dry fiber spools (short spools from fiber production, residual spools from processing) are often crushed to a certain particle size in several successive stages. Mechanical shredding and separation is a simple and technically mature process that has long been used for other materials (e. g. stone, concrete).

CFRP components (matrix-containing material) are cut to a manageable size with the aid of saws or rice mills where necessary, e. g. for very large components. The material is then further processed by means of a crusher or a slow-running cutting mill, where it is crushed to lengths of, for example, 50 to 100 mm. Depending on the fiber length required in the recycled material, a further fine comminution to lengths of 10 to 50 mm is carried out with the aid of hammer mills. After these shredding steps, numerous secondary components still adhere to the fibers. Therefore, a coarse, fiber-rich fraction must usually be separated from a fine, resin-rich fraction with the aid of sieves and cyclones. In the case of dry fiber as the starting material, this step is not necessary. In Figure 6.16, the procedure is shown schematically.

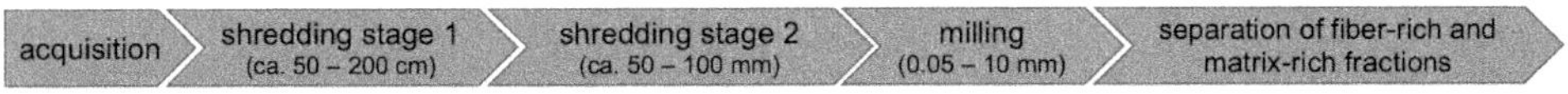

Figure 6.16 Schematic sequence of mechanical shredding

Short spools from carbon fiber production, which are inevitably produced in the production process, often cannot be marketed as such directly by the manufacturers. For this reason, these dry fiber spools (see above) are already cut into short fibers of different lengths or processed into fiber flour by the fiber manufacturer using rotary knives. Although this is not recycling in the traditional sense, as these carbon fibers have never been processed before, it leads to a significant reduction in the amount of waste. This represents the first step in the recycling of carbon fibers and increases the cost-effectiveness of these products.

6.4.2 Pyrolysis Process

Pyrolysis or pyrolytic decomposition is a thermo-chemical decomposition of organic compounds. High temperatures of 200 to 900 °C force covalent bonds to break within large molecules, resulting in smaller molecule chains (depolymerization). In contrast to gasification and combustion, this process takes place exclusively under the influence of high temperatures and in an oxygen-free inert gas atmosphere, e. g. nitrogen. This produces gases, liquids, and solids whose quantity distribution and composition depend both on the starting material and on the pyrolysis temperature and treatment duration. Long-chain as well as short-chain reaction products can be formed, e. g. in the pyrolysis of polymers. Here the corresponding monomer as well as a considerable amount of pyrolysis gas consisting of hydrogen, carbon monoxide, and carbon dioxide as well as methane and higher hydrocarbons are often produced.

For the thermal recycling of waste containing carbon fibers, conventional hot air pyrolysis and microwave pyrolysis are available as industrially applicable processes, as shown in Figure 6.4. In both processes, the long-chain polymer chains of the matrix resin are degraded to short-chain, mostly gaseous hydrocarbons. This resulting pyrolysis gas is utilized via combustion as an additional energy source to generate the necessary process temperatures. At the same time, no dangerous fumes or substances are released into the environment. Typically, the starting material is subjected to temperatures of around 500 to 850 °C to decompose the organic matrix (thermoplastic or thermoset matrix). Figure 6.17 shows an example of the principle process for pyrolytic recycling of carbon fibers. The waste containing carbon fibers is pyrolyzed in the reactor for several hours under exclusion of oxygen and at high temperatures after crushing. However, before the material can be fed into the pyrolysis, it must be crushed in order to keep the residence time of the material in the pyrolysis as short as possible. This enables economical process times and prevents the property level of the carbon fibers from being reduced too much. At the same time, an approximate, later fiber length can be pre-set depending on the application or further processing by means of shredding.

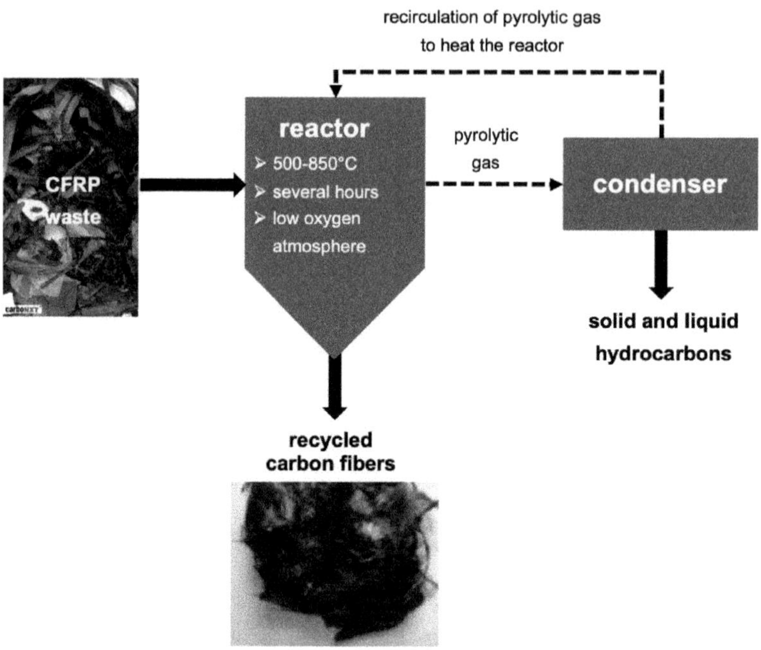

Figure 6.17 Schematic sequence of a pyrolysis [6-4] [6-7]

The quality of the exposed fiber and the degree of degradation of the matrix depend on the temperature applied and the residence time of the material in the pyrolysis chamber. In addition, instead of working under a normal atmosphere, it is also possible to work under protective gas in order to prevent excessive oxidative attack and damage to the fiber surface. Figure 6.18, left, shows the image of a residue-free pyrolyzed fiber. If pyrolysis is inadequate, matrix residues or coke-like residues (pyrolysis coke) can be detected on the fiber surface. A too-strong oxidative attack leads to holes in the filament surfaces and drastically reduces the fiber properties (Figure 6.18, right). Both of these factors worsen the integration of the rCF in a later, new application. In an optimized pyrolysis process, fiber qualities with up to 95% of the original properties such as mechanical strength and fiber diameter can be recovered. However, this requires sufficient pre-shredding so that only fibers with a shorter fiber length can be recovered. Pyrolysis is currently the only commercially operated thermal process for the recovery of carbon fibers.

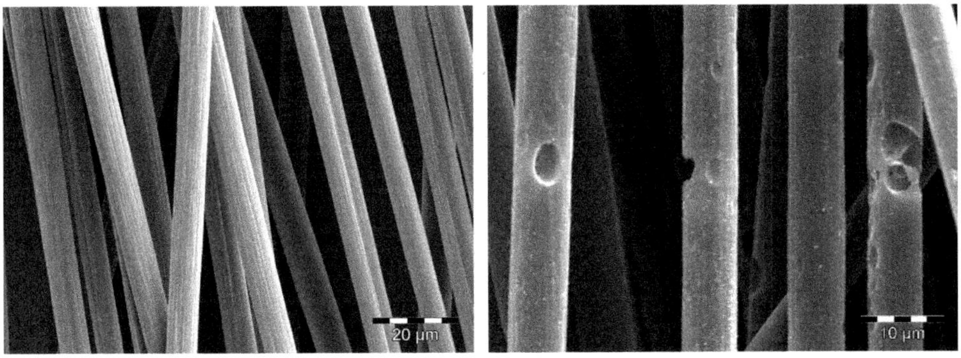

Figure 6.18 Scanning electron microscope images of pyrolyzed carbon fiber [*CarboNXT GmbH*]

6.4.3 Fluidized Bed Process

The fluidized bed process comes from the field of thermal-mechanical process engineering. A fluidized bed is a bed of solid particles that is fluidized by the upward flow of a fluid (liquid or gas) and thus has fluid-like properties. In the fluidized bed, the fluidized material (= solid particles from crushed CFRP waste) comes into close contact with the fluidized medium flowing in from below through the fluidized bed, in this case hot air. In the scheme shown in Figure 6.19, the matrix polymer of the CFRP is depolymerized by the intensive heat exchange with the hot gas and removed from the fluidized bed by the continuous gas flow together with the exposed fibers. This process is a gasification and not a pyrolysis, since the organic matrix material is converted with atmospheric oxygen to carbon monoxide and other fission products, so that the perfect separation from the fiber can be achieved. The recycled fibers and the reaction gases are separated from each other by means of a cyclone. While the rC fibers collect as piles at the bottom of the cyclone, the gas mixture of carbon monoxide, air, and other components is thermally post-treated and the resulting waste heat is used to heat the reactor supply air [6–8].

In contrast to the gasification process described above, fluidized bed pyrolysis, also known as the hamburger process, operates under exclusion of oxygen. This continuously operating process variant is used for chemical recycling, e. g. of plastics and biomaterials. The aim is the recovery of valuable hydrocarbons, which can be used as raw materials in the chemical or petrochemical industry.

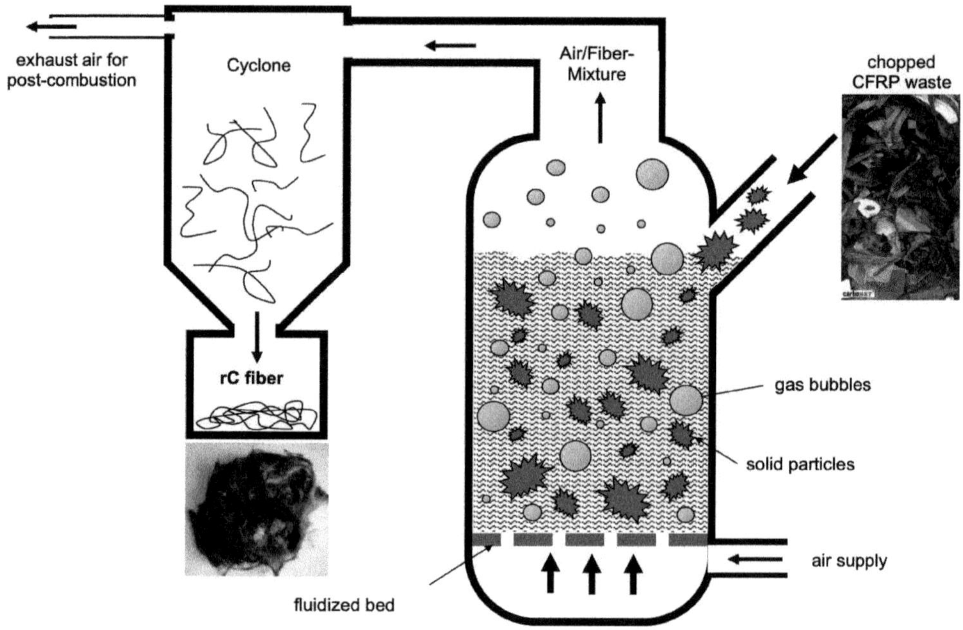

Figure 6.19 Schematic sequence of the fluidized bed process (according to [6-8])

6.4.4 Solvolysis

Solvolysis generally refers to a chemical reaction between a solvent and another reaction partner. The chemical mechanism corresponds to a nucleophilic substitution or elimination reaction in which chemical bonds are broken. The solvolytic reaction is specifically named after the solvent used:

- alcoholysis
- ammonolysis
- hydrolysis

In the chemical recycling of matrix-contaminated carbon fibers (CFRP, prepreg residues), the organic matrix polymer (thermoplastic or thermoset) is depolymerized (decomposed) and liquefied with the aid of the solvent in a chemical reactor into short-chain fragments (monomers or petrochemical raw materials), while the carbon fiber remains as an insoluble component, and can be separated and reprocessed. Depending on the chemical structure of the matrix polymer, the correct solvent must be selected. The liquefied organic material (hydrocarbons) can be used again for the production of polymers or other chemical substances. Before the chemical treatment, the starting material must be mechanically comminuted to increase the specific surface area (improvement of reactivity) and to adapt it to the reactor size – as is necessary during pyrolysis.

One advantage of chemical recycling over thermal recycling is the significantly lower temperatures ($T < 450$ °C) to which the carbon fibers are exposed. By using the right solvent, thermal-chemical damage and thus a reduction in the mechanical performance of the fibers can be avoided. As with thermal recycling, the carbon fibers are completely freed from matrix residues and sizing. Due to different solvents, temperatures, pressures, and other parameters, there is a wide variety of processes in solvolysis. Figure 6.20 shows the schematic sequence of the procedure.

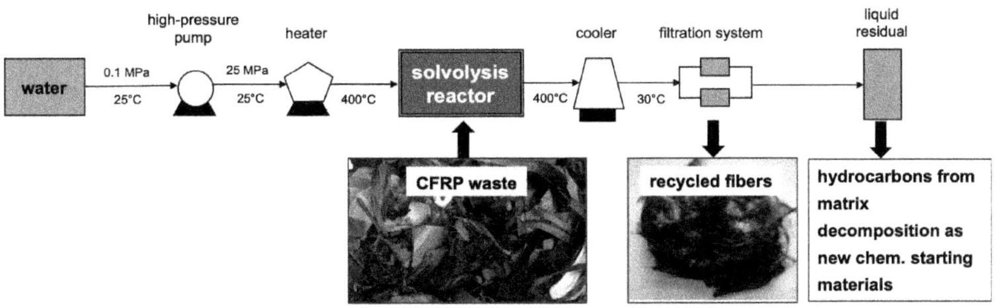

Figure 6.20 Schematic sequence of a classical solvolysis

In addition to a classic solvolysis process, solvolysis with supercritical fluids can also be used. Supercritical fluids are characterized by a high diffusivity and thus have a high diffusion coefficient. This enables a high mass transfer coefficient and thus fast mass transfer processes [6-14]. On the other hand, supercritical fluids are characterized by a high dissolving capacity both for ionic and polar or non-polar bonds, as typically exist in plastics. Compared to classical solvolysis, shorter process times can be achieved and water can be used as a supercritical fluid instead of toxic organic solvents. However, in order to reach the supercritical state, a significantly higher pressure and a higher temperature are necessary, which means a higher energy input.

6.4.5 Chemical Direct Use

Another possibility for the reuse of plastics and CFRP is the use of these materials in a direct chemical transformation for the production of calcium carbide, an important chemical basic material, e. g. for fertilizer production and the production of pig iron. Carbide synthesis in an electric arc furnace requires large amounts of coke and coal as carbon suppliers as well as calcium oxide. The reaction equation underlying carbide synthesis is as follows:

$$3\,C + CaO_{(melt)} \rightarrow CaC_{2\,(melt)} + CO$$

Calcium oxide, also known as quicklime, reacts with carbon to form calcium carbide. The reaction by-product is carbide furnace gas, which consists mainly of carbon monoxide (CO) and hydrogen (H_2). The carbide furnace gas is used for the production of other chemical raw materials. As simple as the chemical process can be described with the equation shown above, its industrial-scale implementation in closed electro-pit furnaces is just as complex. A high energy input is necessary for the production of carbide. In order to generate reaction temperatures of more than 2000 °C, currents of up to 140,000 amperes flow between the electrodes. To produce one metric ton of carbide from about 1000 kg quicklime and about 600 kg coke and coal (black matter), about 3500 kWh of energy is needed.

About 10 years ago, the Southern German chemical company *AlzChem* developed a new process that makes it possible to replace some of the black materials used – carbon and coke – with plastics and carbon fiber-containing waste as secondary raw materials. In this plant, around 1000 metric tons of plastics is to be directly recycled each month as carbon carriers in shredded or pre-pelletized form from industrial and commercial plastics processing. It is important for the process that the quality of the waste is largely constant, i. e. that the plastics are as pure as possible, in order to be able to operate the furnace process with a high carbide yield. A direct competitor to this is the purely thermal recycling of waste. With this substitution of coal and coke by plastics and materials containing carbon fibers, the carbon contained in the plastic products is not only used thermally, but also materially for new products [6-15]. However, recovery of the fiber as such does not take place.

6.4.6 Electrodynamic Fragmentation

Electrodynamic fragmentation is based on the principle that the dielectric strength of a material against electrical discharge is not a physical constant, but is dependent on the pulse duration of the electric discharge and thus inversely proportional to the electrical conductivity of many materials.

This physical relationship makes it possible to introduce an electrical impulse into a solid in a controlled manner. For this purpose, the CFRP solid is positioned in a water bath between two electrodes and subjected to an electrical discharge of 30 to 45 kV with a very short pulse duration (< 500 ns). The pulse energy is about 600 J. As can be seen in Figure 6.21, left, the dielectric strength of water at ultra-short pulses is higher than that of other materials, such as oil or solids. For solids and CFRP, the electrical impulse from one electrode to another therefore does not pass through the surrounding water bath, but along phase boundaries between fiber and matrix polymer, which represent the path of least resistance, through the solid. At the moment of the breakthrough between the electrodes, a plasma channel is formed whose expansion in the solid produces an explosion with a pressure wave of around 9 to 10 GPa and a short-term temperature of up to 10^4 K (see Figure 6.21, right). The fiber composite material can thus be selectively and phase-purely separated along its grain boundaries by introducing several pulsed discharges. This is supported by the dielectric properties of the matrix and the electrically conductive carbon fiber, which results in partial current conduction in the composite material. This is a major difference and advantage over mechanical comminution (crushing, grinding). As a rule, the fibers obtained show only little damage. However, fibers that have already been released must be removed from the process zone of fragmentation, or otherwise they will be shredded further and further and ultimately remain as carbon dust.

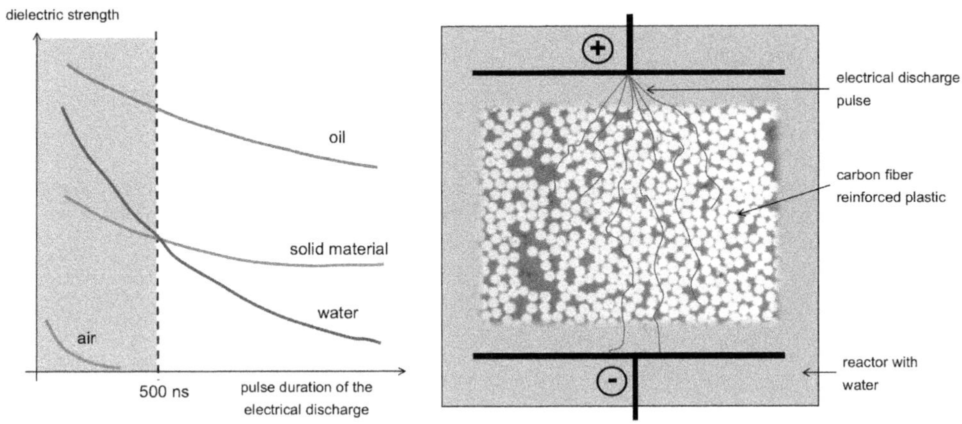

Figure 6.21 Left: dielectric strength of various materials; right: principle of electrodynamic fragmentation (according to [6-16])

6.4.7 New Processes

Further processes are currently being developed to separate the fiber and matrix more cost-effectively and quickly and to remove the matrix from the fiber even more gently. The latter has considerable effects on the mechanical property level of the fiber, which is reduced by strong thermal stress and oxidative attack.

A possible new approach in the field of thermal recycling (pyrolysis) is inductive heating. An alternating electromagnetic field is generated by means of an induction coil, which induces eddy currents in the CFRP material due to the electrical conductivity of the carbon fibers. This results in a direct heating of the fibers by Joule heating (direct heating of the resin by dielectric heating). First tests show a lower energy requirement than with today's pyrolysis processes and very short process times of less than 60 seconds. The results also show that mechanical properties such as tensile strength and stiffness can be 95% similar to those of a new fiber.

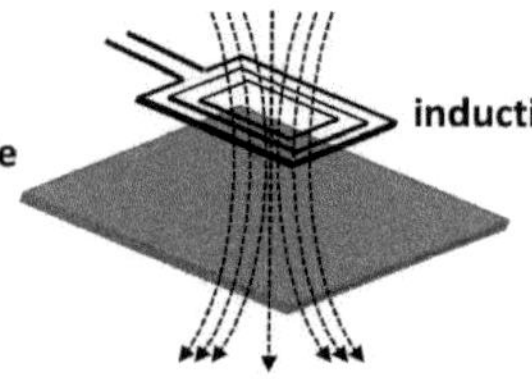

Figure 6.22
Principle of inductive heating of CFRP laminates (according to [6-14])

6.4.8 Overview and Comparison of Procedures

Table 6.5 provides an overview of the separation processes described above and lists examples of the advantages and disadvantages of the processes.

Table 6.5 Overview of Separation Processes (According to [6-14])

Process	Advantages	Drawbacks
Mechanical separation	▪ simple, established process ▪ fiber and matrix recovery ▪ no use or development of substances harmful to health	▪ moderate mechanical properties of rC fibers ▪ unstructured, coarse fiber architecture ▪ limited possibilities for further processing
Pyrolysis	▪ mechanical properties rCF at 95% output fiber level ▪ energy recovery from matrix ▪ almost no residues on fiber surface ▪ rC fiber surface exhibits good adhesion to new matrix polymers ▪ the only industrialized process	▪ residues on fiber surface possible ▪ fiber quality strongly dependent on process control ▪ approx. 5% loss in fiber strength ▪ pyrolysis gas treatment necessary
Solvolysis	▪ mechanical properties rCF at 95% output fiber level ▪ hardly any residues on fiber surface ▪ high potential for recovery of matrix components	▪ long response times typical ▪ use or development of chemical substances harmful to health links ▪ approx. 3% loss in fiber strength ▪ sensitive to impurities in the process ▪ limited scalability

Process	Advantages	Drawbacks
Fluidized bed process	▪ insensitive to impurities (material mix) ▪ known and well-described procedure ▪ hardly any residues on fiber surface	▪ significant fiber length reduction ▪ approx. 25 to 50% loss in fiber strength ▪ residues on fiber surface ▪ no recovery of matrix components possible
Electrodynamic fragmentation	▪ hardly any residues on fiber surface ▪ known and well-described procedure ▪ recovery of matrix polymer possible	▪ limited scalability ▪ fiber length reduction ▪ sensitive to impurities in the process

In summary, it can be stated that the fiber stiffness for rCF is hardly influenced by all processes. The electrical conductivity of the rC fibers compared to a new fiber does not change. In contrast, some processes cause a significant degradation of the fiber strength and lead to damage in the fiber surface, depending on the process control.

6.4.9 Hazards and Health Aspects

In the following, two possible hazards are considered which may arise from the release of particle dusts and fibers during recycling, e.g. during mechanical processing in the course of component manufacture [6-24] [6-25].

Exposure through Inhalation

Depending on the process, the process parameters, and the process- and material-specific properties, as well as the type of carbon fiber processed, matrix and fiber dusts can occur. Thermoset plastics are particularly prone to matrix dusts. The fiber geometry and the fracture behavior are decisive for a possible health hazard. According to current knowledge, carbon fibers do not break or splice longitudinally in conventional mechanical processing methods. Thus carbon fibers differ fundamentally in their behavior compared to asbestos fibers, which show exactly this undesirable splicing behavior. Nevertheless, the nature of the fibers and fiber fragments released during the processing of materials containing carbon fibers should be analytically verified. According to WHO criteria, fibers with a length greater than 5 µm, a diameter less than 3 µm, and a length to diameter ratio (l/d ratio) greater than 3:1 are particularly critical because of their effect on lung tissue. Carbon fibers with diameters of 5 to 7 µm are typically used in CFRP production and recycling. Therefore, the formation of respirable fiber fragments or particles according to the WHO definition is unlikely. More critical are fiber fragments and particles that could be released from thermal processing (pyrolysis, solvolysis, etc.), especially under the influence of oxygen. Field fire tests show an increased occurrence of WHO fibers at temperatures above 650 °C due to decomposition and degradation processes. In the event of an acute fire on CFRP structures, it can be assumed that effective respiratory protection must be used anyway to fight the fire due to the smoke gases produced by the plastic matrix. Thus, there is hardly any danger from possible inhalation of respirable fiber fragments. Appropriate breathing protection measures after the fire are of course necessary.

Electrical Hazard

The dusts, fiber balls, and angel hair from recycling processes that are produced during processing are generally excellent electrical conductors. Suitable protective measures, such as filters, must therefore be provided for electrical and electronic systems in order to avoid short circuits. At the same time, classical electrostatic separators are unsuitable for exhaust gas cleaning due to the electrical conductivity of the fibers.

■ 6.5 rCF Products and Applications

Since the beginning of the recycling of carbon fibers, a large number of different semi-finished recycled products (rCF semi-finished products) have been developed and the development continues steadily. The focus is increasingly on adapting rCF semi-finished products to the requirements of new applications and on the further development and optimization of processes for the manufacture of rCF semi-finished products.

Probably the simplest rCF semi-finished products, which are often also offered by the fiber manufacturers themselves from short spools and cut remnants, are short cut fibers (Figure 6.23) with different fiber lengths as well as milled fibers and granules (Figure 6.24, left).

Milled fibers with fiber lengths of $l < 0.5$ mm or short fibers with $l > 0.5$ mm are suitable for injection molding applications. In addition to pure rCF as flour or short fiber, bound milled fibers (fiber balls) and granulates already compounded with thermoplastics are also available. These rCF semi-finished products are typically used in injection molding applications. The binding of milled CF (fiber balls) serves to reduce the development of CF dust and to improve the flowability of the material.

Figure 6.23 Left: rCF short cut (made of dry fiber); right: rCP short cut (made of prepreg)

Figure 6.24 Left: rCF flour; center: rCF fiber balls; right: rCF/PA granules
[CarboNXT GmbH, Tejin Carbon Europe GmbH]

rCF nonwovens, also known as carbon paper, in low grammages of approximately 10 to 200 g/m^3, can be produced from fiber balls with longer fiber lengths of l = 5 to 15 mm using the wet or airlaid process. For larger fiber lengths of l = 15 to 50 mm, random non-wovens with grammages of approximately 200 to 1200 g/m^3 can be produced.

Figure 6.25 rCF fleece, left: 30 g/m^3 wet-laid fleece; right: 400 g/m^3 needle fleece
[CarboNXT GmbH]

Both types of fleece can be processed into SMC mass by the addition of reaction resins or into organic sheets in the case of thermoplastics. If the fibers or fiber bundles are in straightened form, BMC press mass can be produced at fiber lengths of approximately 20 to 100 mm with the addition of reaction resins. Both semi-finished products are ideally suited, for example, for the production of CFRP plates and assembly clips. The former can be machined in milling processes and used quasi as a metal replacement, e. g. in mechanical engineering.

Figure 6.26 Left: rCF-SMC and machined SMC block; right: BMC ground [*CarboNXT GmbH*]

In textile processes, fiber lengths $l > 50$ mm can bespun into fiber bundles (yarn, rovings), into staple fiber yarns, and further processed into relatively coarse, flat textile semi-finished products with directed properties. A further possibility of new application is the production of a hybrid staple fiber yarn by mixing rC fibers with thermoplastic fibers in a textile opening unit. The fiber flakes opened and mixed in this way are processed into a fiber sliver on a carding machine with the aid of carding processes. To improve filament orientation, the slivers can be stretched in the axial direction in a stretching field and wrapped with a thermoplastic filament, such as polyamide, for stabilization and wound up. The thermoplastic fibers in this hybrid yarn later serve as a matrix, so that the material can be consolidated into one component by hot pressing [6-26] [6-27] [6-28] [6-29].

Figure 6.27 Left: rC staple fiber tape – fully impregnated thermoplastic tape made of recycled carbon fibers and polyamide 6; right: yarn bobbins with staple fiber hybrid yarn made of recycled carbon and polyamide 6 fibers [*DITF Denkendorf*]

The examples presented give a small insight into the range of possible semi-finished rCF products that can be processed into new products and applications for a wide variety of industrial sectors. Figure 6.28 uses a schematic rough outline to show once again the various ways in which carbon fibers can bE recycled depending on the input variables, as well as possible semi-finished rCF products and areas of application that can be produced from them. For better clarification, Figure 6.29 provides examples of application-relevant material properties that are based on the special properties of carbon fibers, which are mirrored in the potential applications of semi-finished rCF products.

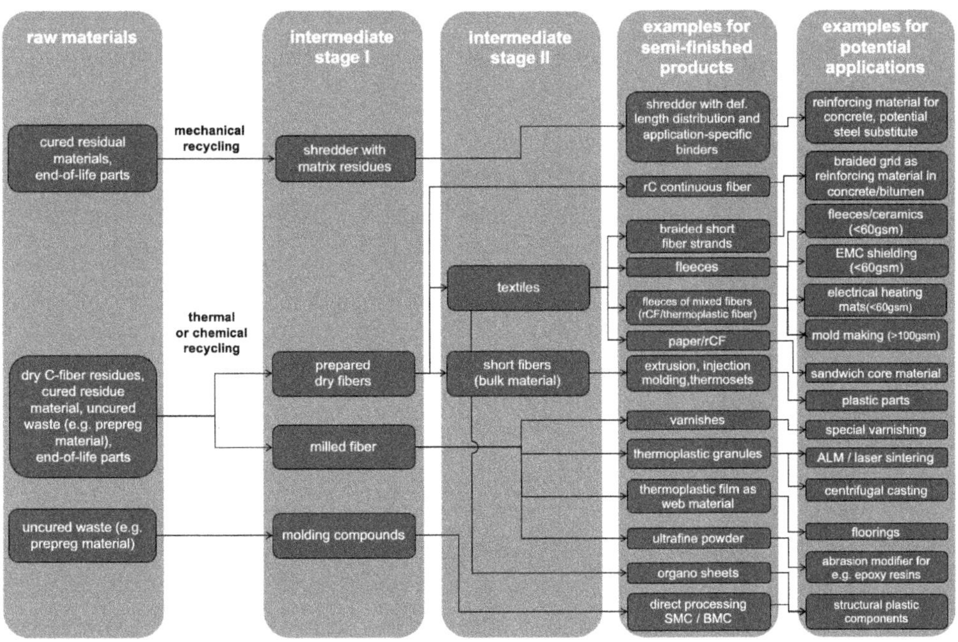

Figure 6.28 Schematic path from raw material to new semi-finished product and applications

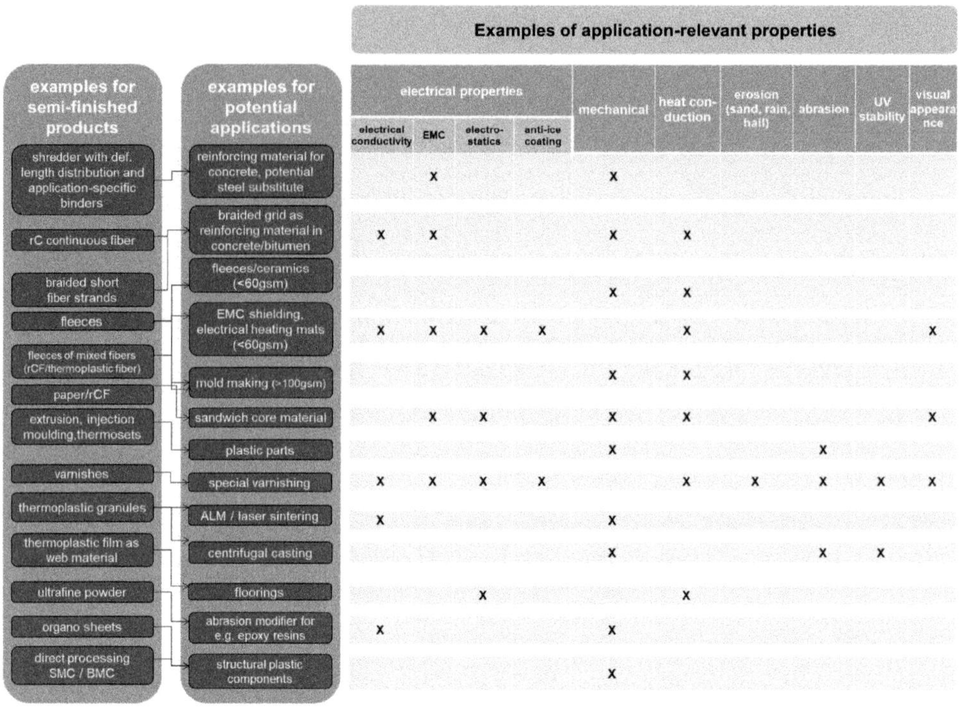

Figure 6.29 Examples of application-relevant properties and potential applications

Both graphics show the great variety of applications, potential as well as existing, for rCF semi-finished products, which are also described in Chapter 5. The applications are not only oriented towards the known good mechanical properties (stiffness, strength) of carbon fibers, but also, for example, electrical properties, heat conduction, or abrasion behavior play a role.

The rCF semi-finished products can offer a comparatively cost-effective and high-quality supplement to new fiber semi-finished products, which are equally suitable for wet pressing processes, compounds, SMC/BMC, prepregs, or thermoplastic organic sheets with a high level of properties. All product examples shown above can thus make an important contribution to the sustainable production and use of carbon fibers in a wide range of applications. This applies all the more if it is possible to create a closed-loop concept so that the carbon fibers are reused several times.

■ 6.6 Producers

On the world market there is a manageable number of recycling companies and suppliers of recycled carbon fibers. Here a clear distinction must be made between:

- "real" recycling companies that process and recycle all types of CFRP components and carbon fiber-containing residues, including auxiliary waste, such as cardboard tubes, foils, and release papers.
- providers who only cut dry fiber remnants (mostly UD fiber spools).

Table 6.6 provides an overview of companies currently on the market, and their range of applications, processes, and derived products.

Table 6.6 Examples of Recycling Companies and rCF Suppliers

Company	Utilization of	Processes	Products
ELG	prepreg, CFRP components, UD dry fiber and fabric waste	pyrolysis, sorting, separating, grinding	rCF nonwovens, SMC, BMC, fiber flour, short fibers
CarboNXT	prepreg, CFRP components, UD dry fibers and fabric waste	pyrolysis, sorting, cutting, grinding, fleece production	rCF nonwovens, SMC, BMC, fiber balls, fiber flour, short fibers, fiber flakes
CAR Fibertec	UD dry fibers	cutting	short fibers
SGL	UD dry fibers	cutting, grinding, fleece production	short fibers, nonwovens
Tenovo	UD dry fibers	cutting, fleece production	short fibers, nonwovens
CFR Industries	prepreg, CFRP components, UD dry fibers and fabric waste	probably: pyrolysis, sorting, cutting	short fibers, fiber flakes
Vartega	UD dry fibers and fabric waste	pyrolysis, sorting, cutting, fleece production	short fibers, nonwovens
CCI - Carbon Conversions	UD dry fibers and fabric waste	sorting, cutting, fleece production	short fibers, nonwovens
Materials Innovation Tech LLC	probably: prepreg, CFRP components, UD dry fibers and fabric waste	- unclear -	probably: short fibers, nonwovens
Karborek Spa	prepreg, CFRP components, UD dry fibers and fabric waste	probably: pyrolysis, sorting, cutting	short fibers, nonwovens
HADEG Recycling GmbH	fabric and scrim waste, UD dry fibers, prepreg	sorting, cutting	reuse of carbon fiber products, short fibers, nonwovens

6.7 Challenges and Trends

The recycling of carbonic waste of all kinds and the associated recovery of high-quality carbon fibers (rCF) on an industrial scale has been solved and is increasingly applied. Due to falling costs due to new or optimized processes along the recycling cycle, a further increase in the distribution of rCF products and applications can be expected due to falling material prices. This is already reflected in the fact that large aircraft manufacturers (OEMs), such as *Airbus* and *Boeing*, enter into cooperation agreements with recycling providers who not only undertake the recycling but also produce new semi-finished products or even components for the OEMs. The biggest obstacle to the use of this valuable secondary raw material at present is primarily the reservations of potential users and OEMs regarding the quality of rCF compared to new fibers. rCF material is often still regarded as supposed class B goods of inferior quality, which represents a considerable obstacle to faster dissemination. This lack of confidence is often due to a lack of information about the material properties required in the CAD design process. Furthermore, to date there has been no uniform standardization of rCF fiber types or uniform quality criteria. On the other hand there are missing specifications on the part of the users.

Outlook

Pioneering work is required in particular in the area of new applications and possible uses for rCF, also beyond the classical mechanical structural reinforcement by carbon fibers. In doing so, the rCF-specific properties must be taken into account and these must be exploited as an advantage in new applications, thus increasing the range of applications. Experience from the past 10 years shows that successful use has always been possible where rCF manufacturers and processors have worked closely together along the entire production chain. Despite the relatively small quantities of rCF produced so far, it can be assumed that the trend towards reuse will continue, as any form of reuse and, above all, multiple use means energy savings compared with a new fiber. At the same time, many efforts are being made towards alternative, non-petrochemical precursors with the aim of making the ecological footprint from this side of the process chain more sustainable (see Chapter 7).

References for Chapter 6

[6-1] M. Kühnel, M. Sauer, Carbon Composites e. V., *The global CF and CC market 2017*, International Composites Congress (ICC) Stuttgart, Sept 2019

[6-2] J. van der Woude, presentation ACMA: *Recycling - Status and Developments in Europe*, Composites Recycling Conference, Apr 10-12, 2018, Knoxville, TX

[6-3] M. von Bistram, Industrielle Verarbeitung von rCF Wertstoffen, SGL Automotive Carbon

Fibers, Workshop Carbon Composite Recycling CCeV Strategiekreis Nachhaltigkeit, Universität Augsburg, Germany, January 2017

[6-4] L. O. Meyer, Untersuchungen zur Pyrolyse. Eine Methode zur Rückgewinnung von Kohlenstofffasern aus CFK-Fertigungsabfällen, Dissertation, Technische Universität Hamburg-Harburg, 2011

[6-5] B. Reckter, *Recycling*, VDI Nachrichten, May 11, 2018, No. 19, p. 23

[6-6] F. Ansorge, *Closing the loop for thermoplastic composites*, JEC Composites Magazine, No. 112, April-May 2017

[6-7] S. Kreibe, B. Hartleitner et al., MAI Recycling - Entwicklung ressourceneffizienter CFK-Recyclingverfahren und Prozessketten für die künftige Bereitstellung qualitativ hochwertiger rC-Halb-zeuge, Förderkennzeichen (support code) 03MAI03A, 2015

[6-8] G. Jiang, S.J. Pickering, G.S. Walker, K.H. Wong, C.D. Rudd, *Surface characterisation of carbon fibre recycled using fluidised bed*, Applied Surface Science, 2008, 254, pp. 2588-2593

[6-9] Siegfried Hauptmann, *Organische Chemie*, 2. durchgesehene Auflage (2nd ed.), VEB Deutscher Verlag für Grundstoffindustrie, Leipzig, 1985, p. 419, ISBN 3-342-00280-8

[6-10] Entry for *Solvolysis* in: IUPAC Compendium of Chemical Terminology (the "Gold Book"). doi:10.1351/goldbook.S05762

[6-11] E. Seiler, U. Teipel, Projektverbund ForCycle, Abschlussbericht (final report): Recycling von Kompositbauteilen aus Kunststoffen als Matrixmaterial - ReKomp, 2017

[6-12] D. Pico, G. Seide, Th. Gries, *Thermo Chemical Processes: Potential Improvement of the Wind Blade Life Cycle*. Chemical Engineering Transactions, 2014, 36, pp. 211-216

[6-13] G. Oliveux, J.L. Bailleul, A. Gillet, O. Mantaux, G. Leeke, *Recovery and Reuse of discontinous carbon fibers by solvolysis: Realignment and properties of remanufactured materials*, Composite and Science Technology, 2017, 139, pp. 99-108

[6-14] A. Schneller, W.M. Mueller, S. Richler, S. Horn, *Recycling of woven carbon fiber patches from laminated CFRP by means of induction heating*, 20th International Conference On Composite Materials, ICCM, Copenhagen, 2015

[6-15] BINE Informationsdienst 08/2011, *http://www.bine.info/publikationen/publikation/carbidmitkunststoffabfaellen-produzieren/*, accessed July 2018

[6-16] H. Bluhm, W. Frey, H. Giese, P. Hoppé, C. Schultheiß, R. Sträßner, *Application of pulsed HV discharges to material fragmentation and recycling*, IEEE Transactions on Dielectrics and Electrical Insulation, 2000, 7(5), pp. 625-635

[6-17] C. Orzol, H.G. Jäckel, *Kohlenstofffaserrecycling, Voruntersuchungen zum elektrohydraulischen Aufschluss von kohlenstofffaserverstärktem Duroplast*, in Recycling und Rohstoffe - Vol. 10, K.J. Thomé-Kozmiensky, D. Goldmann, Eds., ISBN 978-3-944310-34-3, 2017, pp. 229-234

[6-18] S. Eisert, A. Henning, Versuchsanlage für die Zerkleinerung von Kohlefaserverbundwerkstoffen. Modifizierte Schockwellenanlage EHF400, Dresden, 2016

[6-19] C. Hostettler, S. Eisert, Intelligente Zerkleinerungstechnologie zur Erschließung von Sekundärrohstoffquellen. Anwendungsgebiete der Schockwellenzerkleinerung, Dresden, 2016

[6-20] S. Seifert, V. Thome, C. Karlstetter, *Elektrodynamische Fragmentierung - Eine Technologie zur effektiven Aufbereitung von Abfallströmen*, in Recycling und Rohstoffe - Vol. 7, K.J. Thomé-Kozmiensky, D. Goldmann, Eds., ISBN 978-3-944310-09-1, 2014, pp. 431-438

[6-21] T. Müller et al., *Method and device for the disintegration of a recyclable item*, WO201360403A1, 2013

[6-22] CleanSky proposal JTI-CS-2012-1-ECO-01-053, *Disintegration of fibre-reinforced composites by electrodynamic fragmentation technique*, 2012

[6-23] D. Meiners, B. Eversmann, *Recycling von Carbonfasern*, in Recycling und Rohstoffe - Vol. 7, K.J. Thomé-Kozmiensky, D. Goldmann, Eds., ISBN 978-3-944310-09-1, 2014, pp. 371-378

[6-24] Bearbeitung von CFK Materialien, DGVU-Information, Fachbereich Holz und Metall, FB HM-074, Issue 10/2014

[6-25] S. Eibl, N. Scholz, *Besondere Gefährdungen beim Abbrand von Carbon-Kunststoffe*, Deutsche Feuerwehr-Zeitung BRANDSCHUTZ, Issue 6/2014, Verlag W. Kohlhammer, Stuttgart

[6-26] T. Rademacker, M. Fette, G. Jüptner, *Nachhaltiger Einsatz von Carbonfasern dank CFK-Recycling*, lightweight.design, 5/2018, pp. 12–19

[6-27] J. Wellekötter, S. Baz, J. Schwingel, C. Bonten, G. T. Gresser, P. Middendorf, *Kohlenstofffasern im Kreislauf halten*, Kunststoffe 9/2018, pp. 101–105

[6-28] C. Goergen, S. Baz, P. Mitschang, G. T. Gresser, *Organobleche aus recycelten Carbon-Stapelfasergarnen*, lightweight.design, 3/2017

[6-29] C. Goergen, S. Baz, P. Mitschang, G. T. Gresser, Recycled Carbon Fibers in Complex Structural Parts – Organic Sheets Made of rCF Staple Fiber Yarns; Key Engineering Materials Vol. 742, 21st Symposium on Composites; pp. 602–609; 2017; doi:10.4028/, *www.scientific.net/KEM.742.602*, accessed July 2018

7 New Developments

As already explained in Chapter 3, there are three main cost drivers involved in the production of carbon fibers:

- precursor costs
- energy costs
- investment costs for plants

As a result of the chemical conversion processes, around 50% of the precursor's initial mass is lost as mainly gaseous or solid tar-like by-products, which additionally increases the precursor's costs. The gaseous products are typically thermal post-treated and the resulting energy is fed back into the process. At the same time, the conversion of the precursor requires considerable amounts of energy (around 700 GJ/t) on special systems that can only be offered by a few companies on the world market.

The aim of new developments is therefore to configure the manufacturing costs of the precursor and the conversion processes as economically and ecologically as possible.

In the field of research and development, there are several promising approaches for new precursors on a different chemical basis in addition to the classical PAN- and pitch-based precursors. Current research projects therefore also focus on precursors which are cost-effective, have a higher carbon content, and can ideally be represented on the basis of renewable raw materials. This addresses the most important of the three cost drivers in carbon fiber production. However, so far it has not been possible to find alternative precursors to successfully commercialize in large quantities.

In addition to material costs, research into new, less energy-intensive processes for converting precursors into carbon fibers is an important area of research for producing these fibers more cost-effectively in the future. In the manufacturing process, the heat treatment of the fibers (stabilization and carbonization) accounts for about 69% of the total energy costs. Of this, 48% alone is consumed by fiber stabilization (Figure 7.1).

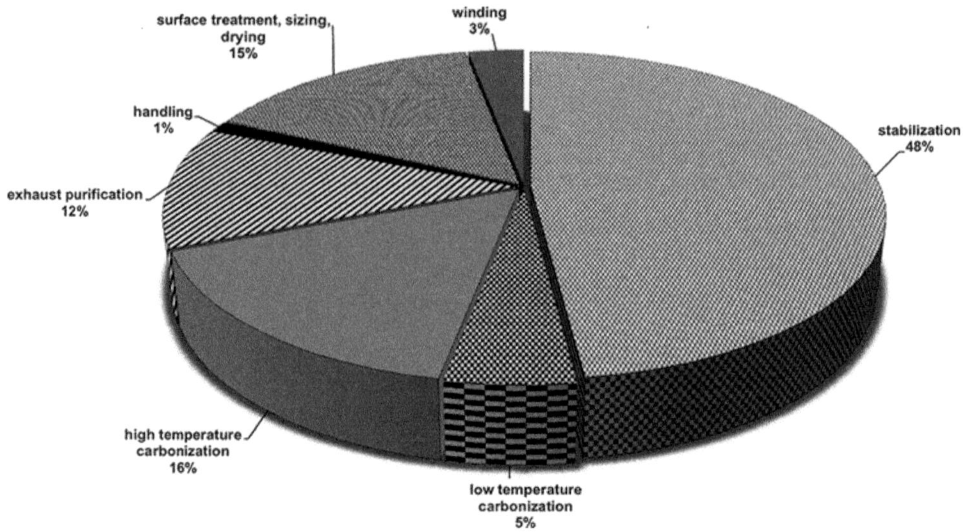

Figure 7.1 Energy consumption in the production process of carbon fibers [7-26]

The stabilization and carbonization thus represent the most time-consuming and energy-intensive step in the carbon fiber manufacturing chain. This means that the highest potential for optimization in terms of cost reductions and positive effects on the environment can be seen in this process stage. In view of the scale of carbon fiber production, even incremental improvements to reduce processing temperatures have an important impact on the overall process and at the global level. The possible positive environmental effects of the use of carbon fibers and their composites are twofold. On the one hand, there is an initial, direct benefit through lower energy consumption during production. On the other hand, the lower material costs resulting from lower manufacturing costs can have a positive impact on the increased use of carbon fibers in light vehicles, which in turn can reduce fuel consumption and vehicle emissions.

In addition to developing new processes or process variants, efforts are concentrated also on looking for alternative locations for carbonization plants where energy costs are significantly lower due to their generation. Here, northern Sweden seems to be the right place, where 100% of electrical energy is generated from wind and water power [7-1].

7.1 Alternative Precursors

Currently potential alternative precursors are:

- polyethylene
- cellulose
- lignin

Research is also being conducted into the use of textile PAN and blends of lignin and PAN.

- textile PAN
- lignin/PAN blends

Table 7.1 gives an overview of the carbon content of the starting material (precursor) and the achievable yield of carbon fiber.

Table 7.1 Raw Material Sources for the Production of Carbon Fibers

Resource	Carbon Content in %	Yield CF in %
PAN	68	45
pitch (MPP)	94	85
polyethylenes	86	75
lignin	62	40
cellulose	45	20-30

7.1.1 Polyethylene

Polyethylene is the most commonly used industrial thermoplastic. Annual production currently exceeds 80 million metric tons per year [7-2]. There are very many different modifications of polyethylene that generally all follow the chemical formula in Figure 7.2. The carbon content of polyethylene is about 86%, which makes this material very interesting as a precursor for carbon fibers.

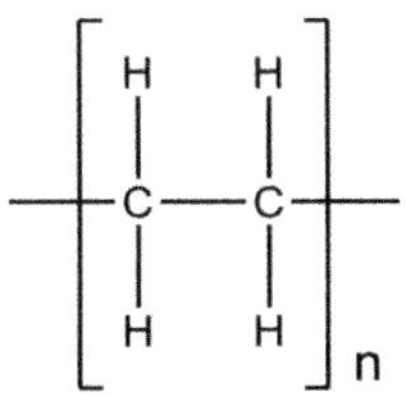

Figure 7.2
Chemical structure of polyethylene [7-2]

However, polyethylene cannot be stabilized in the conventional oxidative stabilization process. An oxidation reaction would decompose the polyethylene precursor and produce water and carbon dioxide as products. For this reason, new stabilization processes had to be developed:

- Treatment with chlorosulfonic acid at 60 to 90 °C
- Treatment with 98% sulfuric acid at 100 to 180 °C

Both methods lead to a higher temperature resistance of the precursor fiber and thus make the carbonization process possible in the first place [7-3].

The mass loss due to carbonization is also considerable with polyethylene - 20 to 50% of the precursor fiber is lost in the process. The biggest problem in the commercialization of polyethylene as a precursor for carbon fibers is the enormous amount of contaminated acid produced as waste [7-4]. Similar to pitch and PAN based precursors, the price of conventional polyethylene is directly linked to the price of crude oil, which is expected to rise further in the future.

7.1.2 Cellulose

Cellulose is the oldest known precursor for the production of carbon fibers. Its use was first documented by Thomas Edison in 1880 during the development of incandescent lamps [7-5]. Since the 1950s, developments based on cellulose have been underway, which are referenced in the literature as rayon. Cellulose has the chemical formula $C_6H_{10}O_5$. The chemical structure is illustrated in Figure 7.3.

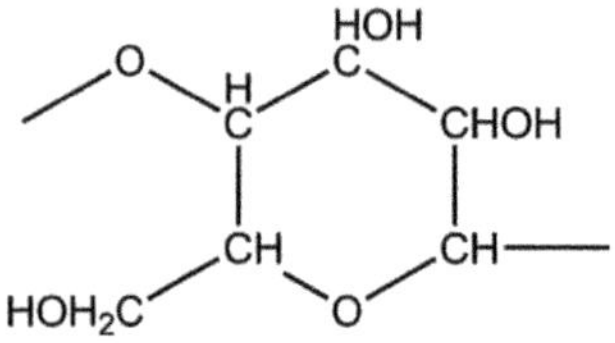

Figure 7.3
Chemical structure the repeating unit of cellulose (rayon)

The specific processes for the production of carbon fibers from cellulose are currently still subject to secrecy on the part of the manufacturers. However, the essential steps to get from insoluble cellulose via soluble and spinnable cellulose to rayon fiber are known. This process is called the viscose-rayon process, the general sequence of which is described in Figure 7.4.

In the first step, the insoluble cellulose is chemically converted into a soluble compound in the stepping process, which is then converted into a viscous, spinnable solution with the aid of a catalyst in the subsequent step (xanthation). This spinning solution can now be spun in a sulfuric acid bath. The rayon fiber is obtained as a natural polymer composed of glucose units connected by ether bridges (Figure 7.5).

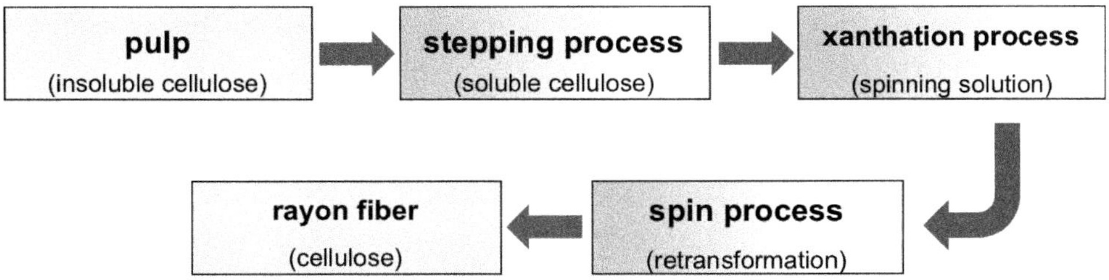

Figure 7.4 Viscose rayon process

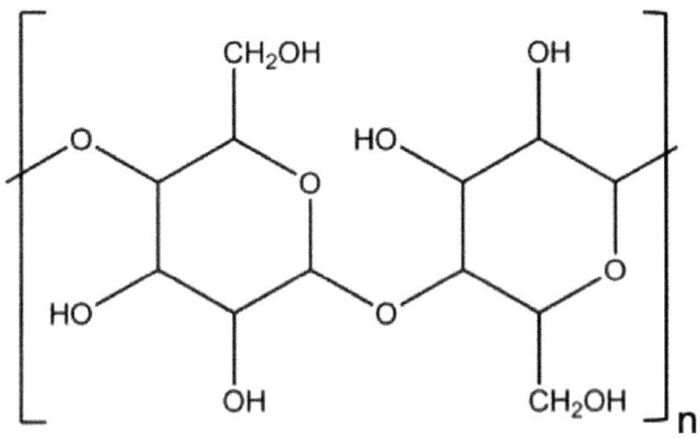

Figure 7.5 Rayon structure

The worldwide production of cellulose amounts to about 14 to 15 million tons per year, at a price of about US$ 1350 per ton. Thus, this raw material is readily available at a relatively low price compared to conventional precursors such as PAN and MPP.

The biggest disadvantage of cellulose as a precursor raw material, however, is its low carbon content of only 45% and simultaneously low fiber yield of 20 to 30% (see Table 7.1). Furthermore, the mechanical properties of rayon-based carbon fibers are rather low compared to PAN-based precursors, so that commercialization is currently not expected [7-6] [7-7].

7.1.3 Lignin

The use of lignin as a precursor is still relatively new and unknown. Lignin combines some of the best economic and ecological properties of all precursor materials available to date. The expected cost advantage and the given sustainability as a natural component of plants and especially wood make lignin the subject of current research, among other things as precursor for cost-efficient automotive carbon fibers [7-8].

Lignin is an amorphous, heterogeneous, and at the same time the second most important natural polymer. As a waste product (essential organic component of black liquor) of the paper industry, it is available in enormous quantities. Since the calorific value of lignin is quite high at around 23.4 MJ/kg, waste has so far been incinerated primarily to generate thermal energy in paper production. Global production of lignin is estimated at over 120 million metric tons per year at a price far below US$ 250 per metric ton [7-9]. In comparison to PAN with manufacturing costs of approx. US$ 2400 per metric ton, cost reductions of over 80% can be shown for the precursor on a lignin basis. As a natural polymer, lignin has a better ecological (CO_2) footprint than PAN, pitch, and polyethylene, which are obtained from crude oil and therefore have a relatively poor CO_2 balance.

Figure 7.6 shows an example of the chemical structure of lignin.

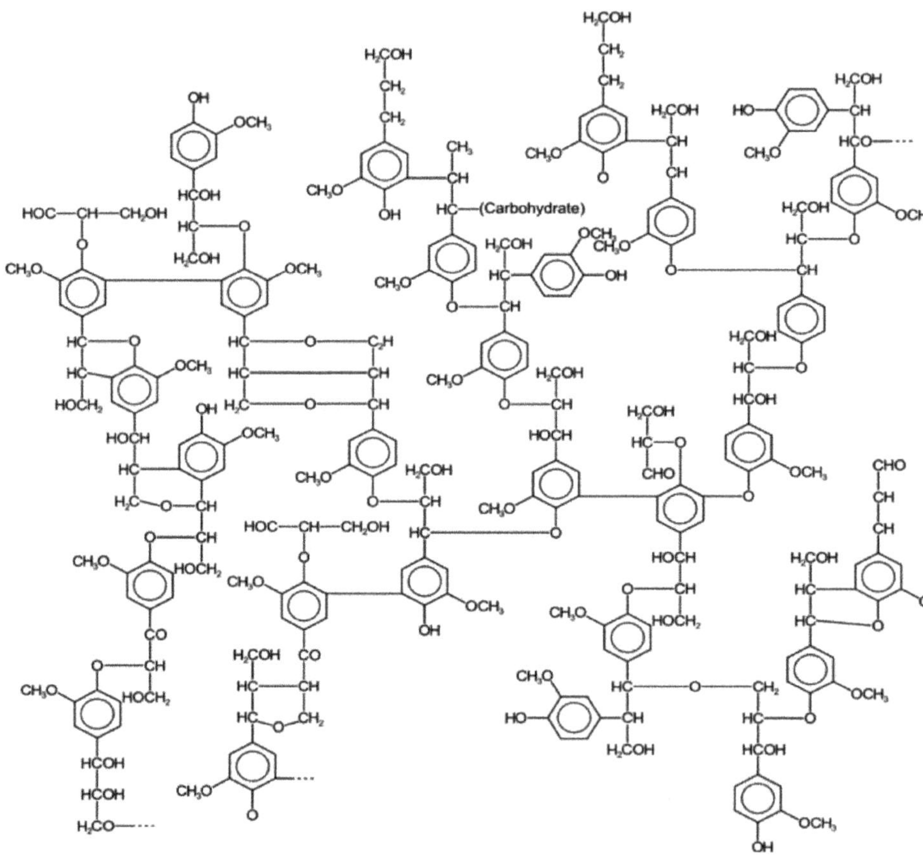

Figure 7.6 Example of the chemical structure of lignin [7-10]

The conventional manufacturing process of a PAN-based carbon fiber comprises three main steps: fiber spinning, fiber stabilization, and carbonization (see Chapter 3). For a future scale-up of carbon fiber production based on lignin, a process must be created that is comparable to today's manufacturing processes. This will help to reduce the investment costs for future production facilities, as conventional fiber lines can be used for the production of lignin carbon fibers

Compared to the conventional PAN-based manufacturing process, two more steps have to be added to prepare the fiber spinning. These two steps are washing and drying as well as pelletizing the lignin powder. This leads to a total of five process steps that enable the use of hardwood lignin as a precursor for carbon fiber production:

- Washing and drying of lignin powder
- Pelletizing the lignin powder
- Melt spinning of the lignin fiber
- Oxidation of the fiber
- Carbonization of the fiber

Figure 7.7 Lignin powder

A lignin powder cannot be used directly for fiber spinning. The main reason is the poor flowability of the powder due to a usually high moisture content (water content: up to approx. 6.1%) as well as volatile components which can accumulate in the hydrophilic lignin powder. The lignin powder must be pelletized in order to improve flow behavior and reduce moisture and volatile components. This can be accomplished for example in an extrusion line.

Figure 7.8
Lignin pellets

The pellets obtained can then be spun into a precursor yarn using the melt spinning process (see Chapter 3). This spinning process is also the most important step in the production of a lignin-based carbon fiber, since the quality of the precursor yarn also determines the quality of the subsequent carbon fiber after carbonization.

Currently, there are still open questions regarding the chemical structure of lignin after extraction of the remaining wood components and re-polymerization in the extrusion process. The lignin polymerizes in a different way than in the natural growth process. Studies show variations in molecular weight, conformation, and some chemical functionalities [7-11]. One reason might be found in the branched and less linear chemical structure of lignin.

The industrial commercialization of lignin-based precursors is being worked on in particular in Sweden, where the paper industry is facing major challenges due to the downward trend in the use of paper as a writing and printing medium. The goal is to develop high-quality products such as lignin-based precursors for carbon fibers by 2025 and to manufacture them industrially, including carbon fibers. At the same time, new applications for carbon fibers are being researched in order to develop a new market [7-12].

7.1.4 Textile PAN

The use of textile PAN (manufacture of clothing) as a precursor for carbon fibers has a significant cost advantage, since the starting materials used are significantly cheaper. Polyacrylonitrile is a terpolymer and is composed of three different monomers by polymerization. Typically acrylonitrile, vinyl acetate, and itaconic acid are polymerized to PAN. Depending on the chemical structure, the monomers are either more or less expensive. Through the variations of this molecular structure either textile PAN or "carbon fiber" PAN can be produced. Depending on the chemical structure of the monomers, they can have, for example, better mechanical properties of the later carbon fiber and/or higher thermal stability in the oxidation process before carbonization to prevent burning of the precursor yarn. This means that when textile PAN is used, both the heat input through the furnace and the dissipation of the resulting reaction heat from the oxidation process must be controlled. This control must be done with an even tighter temperature regulation than with "normal" PAN (see also Chapter 3). This is all the more true with high filament counts of up to 450 K in the yarn bundle.

As mentioned above, a change to a textile PAN, like the high K number, has an effect on the properties of the later carbon fiber. Nevertheless the properties are in general sufficient to be used as short fiber material for SMC/BMC semi-finished products, for example in the automotive industry. *LeMond*, USA, has successfully demonstrated that this type of process control is possible by manufacturing a 400 K roving based on textile PAN (fiber type: GRAIL™, see also Chapter 3). Steps are currently being taken towards further industrialization and commercialization [7-13] [7-14] [7-15].

7.1.5 Lignin/PAN Blends

Another way to reduce precursor costs is to produce a blend of PAN and lignin. The aim is for the lignin to reduce the proportion of high-priced PAN without adversely affecting the thermal and mechanical properties of the precursor and the resulting fiber.

Research results show that although both components can be mixed relatively homogeneously, the mechanical properties of the carbon fibers resulting from these precursors are still significantly lower than those of comparable purely PAN-based fibers. Thus, for example, in a 50:50 blend PAN lignin, a tensile strength of 1200 MPa and a tensile modulus of only 130 GPa were found. This means that the strength is about 75% and the modulus of elasticity about 50% lower compared to a commercial PAN-based carbon fiber. The difficulty in the development and carbonization of such new precursors is often that the quantities obtained are sufficient for chemical-physical characterization but not for extensive mechanical testing, so that little data is available on the mechanical performance of these fibers [7-16] [7-17] [7-18].

7.1.6 Nanofibers and Carbon Nanotubes

The use of nanofibers based on different raw materials (PAN, lignin, blends) or carbon nanotubes (CNTs) for precursors is being investigated in the latest developments. In 2018, a Chinese research team from Tsinghua University in Beijing succeeded developing a process that could produce fibers several centimeters long from bundles of carbon nanotubes. Their tensile strengths should be 9 to 45 times higher (50 to 80 GPa) than the commercially available carbon fibers (3 to 6 GPa), with a very low fiber density of around 1.6 g/cm^3. According to the developers, the fibers should be able to carry a weight of about 800 tons. The future application is imagined as a suspension rope for a space elevator. The tensile strength of such a rope would have to be so high over a length of approximately 100,000 km that it would not tear under its own weight and could carry a transport gondola [7-19] [7-20].

■ 7.2 Alternative Processes

As mentioned earlier, new processes focus on technologies that can replace today's electrically heated oxidation and carbonization furnaces or that support today's process technologies in terms of reducing energy consumption. The stabilization of the PAN precursor is the first critical step in which the yarn undergoes a series of thermally induced transformations and is stabilized for subsequent carbonization. Current technologies in this area are, for example, the microwave-assisted plasma oxidation of the precursor, the heating of the carbonation furnaces by means of natural gas burners, or the use of special additives on the precursor as suitable energy-saving measures.

7.2.1 Plasma Oxidation

4M, USA, is about to commercialize the world's first industrial scale carbon fiber plasma oxidation furnace using atmospheric pressure plasma. Microwaves are often used to support plasma formation. This technology is intended to enable the production of carbon fibers with outstanding properties at significantly lower costs. This is intended to provide the steadily growing fiber composite industry with a cost-effective opportunity to increase its production capacity by up to 300% over the next 10 years. This can significantly reduce market growth costs for industrial carbon fibers in existing markets and promote their use for new applications. Plasma oxidation technology is said to triple production throughput while reducing energy consumption by up to 75% [7-21] [7-22] [7-23] [7-24] [7-25]:

- Acceleration of the oxidation rate for PAN (up to factor 3) and for lignin (up to factor 5)
- Energy consumption less than 25% compared to conventional process (75% savings)
- Up to three times shorter furnace length required - less space required in a conversion line
- 20% reduction in the overall carbon fiber manufacturing process from the above benefits
- Generation of thermal energy in the immediate vicinity of the precursor and uniform distribution on the precursor yarn
- Better control of energy release through oxidative conversion of precursor

Furthermore, there are numerous, further investigations of different research groups on microwave-supported plasma technology and the use of lasers as an energy source, so that further progress can be expected in the coming years [7-27] [7-28] [7-29] [7-30].

7.2.2 Use of Additives

A further research focus for the reduction of temperatures in the stabilization and carbonization process (and thus of the energy requirement) is the use of ionic liquids. Ionic liquids are ideal for impregnating PAN fibers because of their negligible vapor pressure, high temperature stability, low flammability, and low toxicity. Ionic liquids can also be easily adapted to a specific application. In the literature, ionic liquids are described as catalysts for a variety of chemical reactions, so that an application of these compounds in connection with the production of carbon fibers is obvious.

In a 2016 study, 1-butyl-3-methylimidazolium acetate (bmimOAc) was used as an ionic liquid to impregnate (to about 3.2% by weight) a PAN precursor [7-26]. The selection was based on the similarity of the acetate anion to the typical carboxylic acid-functionalized co-monomers used in PAN production. These co-monomers also help to reduce process temperatures and control exothermic conversion reactions. In the study [7-26], the process temperatures could be reduced significantly over the different heating zones, e.g. in zone 1 of stabilization by up to 45 °C (see also Chapter 3), which led to significant energy savings. While the high costs for ionic liquids could potentially limit the cost reduction on the energy side, the relatively small amount absorbed by the precursor of approximately 3% by weight contrasts with this, so that their application appears to be sensible at present.

7.2.3 Natural Gas Burner

The current second option for heating the carbonization furnaces is natural gas burners. The variant is offered by *Harper International,* USA, in combination with conventional electric heating. Like the electric heaters, the natural gas burners offer temperatures of up to 2800 °C, but at a cost about three times lower than electricity. Hybrid furnaces are offered for this purpose, the advantages of which are not only the operating costs but also the flexibility of the energy supply, lower investment costs, and shorter delivery times [7-31].

7.3 Outlook

Carbon fibers and products made from them, such as fiber-reinforced composites, comprise an established material class of their own which, with their unique properties, have become indispensable in many areas of technology in today's world. Whereas the development and manufacture of such fibers has so far often focused on good mechanical properties in order to be able to use the fibers universally as a substitute for metallic materials, the trend is towards large quantities of cheaper carbon fibers, whose mechanical properties may well be at a different, lower level than at present, e.g. for aeronautical applications. This approach is being pursued in particular by the automotive industry in order to reduce costs for CFRP components and meet the demands for an improved ecological footprint.

References for Chapter 7

[7-1] *Carbon fiber production in the north of Sweden,* JEC Composites Magazine, 10/2018, 124, p. 17

[7-2] J. Holbery, D. Houston, *Natural-fiber-reinforced polymer composites in automotive applications.* JOM, 2006, 58, pp. 80-86

[7-3] I. Karacan, K. Ş. Tunçel, *Thermal stabilization of poly(hexamethylene adipamide) fibers in the presence of ferric chloride prior to carbonization.* Polymer Degradation and Stability, 2013, 98, pp. 1869-1881

[7-4] D. Zhang, Q. Sun, *Structure and properties development during the conversion of polyethylene precursors to carbon fibers.* Journal of Applied Polymer Science, 1996, 62, pp. 367-373

[7-5] T. Edison, US 470925, 1880-03-15

[7-6] K. Kong, L. Deng, I. A. Kinloch, R. Young, *Production of carbon fibers from a pyrolysed and graphitised liquid crystalline cellulose fiber precursor,* Journal of Material Science, 2012, 47, pp. 5402-5410

[7-7] G. Goldhalm, *Tencel Carbon Precursor,* Lenzinger Berichte, 2012, 90, pp. 58-63

[7-8] H. Mainka, O. Tager, E. Korner, L. Hilfert, S. Busse, F. T. Edelmann, A. S. Herrmann, *Lignin - an alternative precursor for sustainable and cost-effective automotive carbon fiber.* Journal of Materials Research and Technology, July-September 2015, 4(3), pp. 283-296

[7-9] H. Mainka, *Lignin as an Alternative Precursor for a Sustainable and Cost-Effective Carbon Fiber for the Automotive Industry* (Science-Report from the Faserinstitut Bremen, Vol. 10) ISBN-13: 978-3832539726; May 2015

[7-10] P. Sitte, E. W. Weiler, J. W. Kadereit, A. Bresinsky, C. Körneror, E. Strasburger, *Strasburger Lehrbuch der Botanik für Hochschulen*; ISBN-13: 978-3827410108, August 2002

[7-11] F. Souto, V. Calado, N. Pereira, *Lignin-based carbon fiber: a current overview*, Material Research Express, 2018, 5(7), Article 072001

[7-12] P. Avegard, P. Omani, I. Hansson, H. Hansson, Road map 2014 to 2025: *Swedish lignin-based carbon fibre in composite materials of the future*, Swerea SICOMP, accessed November 2018: *http://www.innventia.com/en/Our-Expertise/New-materials/Carbon-fibres/Swedish-lignin-based-carbon-fibre/*

[7-13] Le Mond: *http://lemond.cc*, retrieval date November 2018

[7-14] Carbonnexus: *https://www.carbonnexus.com.au*, release date November 2018

[7-15] M. Özgür Seydibeyoglu, *A novel partially biobased PAN-Lignin blend as a potential carbon fiber precursor*, Journal of Biomedicine and Biotechnology, 2012, Article ID598324, *doi:10.1155/2012/598324*

[7-16] J. Jin, A. Ogale, *Carbon fibers derived from wet-spinning of equi-component lignin/polyacrylonitrile blends*, Journal of Applied Polymer Science, 2018, 135, 45903

[7-17] G. Ramasubramanian, *Influence of Lignin modification on PAN-Lignin copolymers as potential carbon fiber precursors* (2013), Graduate Theses and Dissertations, 13438, Iowa State University, *https://lib.dr.iastate.edu/etd/13438*, accessed November 2018

[7-18] C. D. Warren, *Development of Low- Cost, High Strength Commercial Textile Precursor (PAN-MA)*, Oak Ridge National Laboratory, 2014

[7-19] Y. Bai, R. Zhang, X. Ye et al., *Carbon nanotube bundles with tensile strength over 80 GPa, Nature Nanotechnology Volume 13*, pp. 589–595 (2018)

[7-20] Golem.de: *https://www.golem.de/news/materialforschung-forscher-stellen-ultrastabile-kohlenstofffasern-her-1810-137403.html*, date of release November 2018

[7-21] 4M Carbon Fiber, *http://4mio.com*, date of issue Oct 31, 2018

[7-22] Patent: *Apparatus and Method for Stabilization and Oxidation of Polymeric Materials*, 11/344,573, US7534854, Jan 31, 2006

[7-23] Patent: *Apparatus and Method for Stabilization or Oxidation of Polymeric Materials*, 11/391,615, US7649078, Mar 28, 2006

[7-24] Patent: *Apparatus and Method for Stabilization and Oxidation of Polymeric Materials*, 12/466,092, US7786253, May 14, 2009

[7-25] Patent: *Atmospheric Pressure Plasma Processing of Polymeric Materials Utilizing Close Proximity Indirect Exposure*, 13/680,406, patent pending, Nov 19, 2012

[7-26] M. Maghe, C. Creighton, L. Henderson, M. G. Huson, S. Nunna, S. Atkiss, N. Byrned, B. L. Fox, *Using ionic liquids to reduce energy consumption for carbon fibre production*, J. Mater. Chem. A, 2016, 4, pp. 16619–16626

[7-27] E. O. Daniels, *Laser Heat Treatment Processing for PAN Based Carbon Fiber Structure Modification*, Florida State University, Master Thesis 2017.

[7-28] L. A. Smith, *Microwave Plasma Surface Modifications of Carbon Fibers*, Master Thesis, Graduate School of Clemson University, 2016

[7-29] X. Luo, *Chemical Characterization of PAN based Carbon Fibers produced by Microwave assisted Plasma (MAP) Technology and Effect of Plasma Treatment on Carbon Fibers Surface and Interphase with Polymer Matrix*, University of Tennessee, Dissertation 2006

[7-30] F. L. Paulauskas, T. S. Bigelow, K. D. Yarborough; *Manufacturing of Carbon Fibers Using Microwave-Assisted Plasma Technology,* Oak Ridge National Laboratory, Oak Ridge, Tennessee, 2000.

[7-31] Company brochures, Harper International, 2018

[7-32] H. Jäger, P. Haider, *Chancen und Strategien zur Bereitstellung alternativer C-Faser-Precursoren,* lightweight.design, 3/2014, pp. 24-28

Index

X

Y